Statistics
A Tutorial Workbook

ROBERT PISANI
UNIVERSITY OF CALIFORNIA · BERKELEY

W · W · NORTON & COMPANY
NEW YORK · LONDON

Published simultaneously in Canada by Penguin Books Canada Ltd.,
2801 John Street, Markham, Ontario L3R 1B4.
Printed in the United States of America.

W. W. Norton & Company, Inc.,
500 Fifth Avenue, New York, N.Y. 10110

W. W. Norton & Company, Ltd.,
37 Great Russell Street, London WC1B 3NU

ISBN 0-393-95457-9

2 3 4 5 6 7 8 9 0

Contents

Preface

This workbook is a companion volume to the text

<u>Statistics</u> by D. Freedman, R. Pisani and R. Purves
(W. W. Norton & Co., New York, 1978)

It is meant to be used in parallel with that text, which is called "FPP" or "the main text" throughout this book. The workbook treats chapter-by-chapter the same topics found in the main text—in some cases it covers additional related material or ideas only suggested in the main text, with the object of deepening the student's understanding. With the exception of the sections on conditional probability in Chapters 14 and 15 of the workbook, none of the additional material should be omitted. Conditional probability may be skipped if the instructor wishes, although the treatment in this workbook makes it especially easy and transparent.

Each chapter is organized as follows. There is a series of instructional sections which outline the major activities of the chapter. Each of these sections contains some worked examples, followed by a number of similar problems which the student should solve using the technique discussed in the section. The answers to all of these problems are found at the end of the chapter. After these sections on major activities comes a longer problem section, called "Exercises". The method of solution is not indicated in this section—if the student has read and worked all the problems in the first series of sections, he or she should be able to recognize the problem type and apply the appropriate solution method. Solutions to odd-numbered Exercises are found at the end of the chapter; solutions to the even-numbered Exercises are found in the <u>Instructor's Manual</u>, so that these problems may be used for homeworks, quizzes and exams.

Acknowledgments

Many thanks to Shel Silver and David Douglas for helping me launch this project, to countless students who have provided questions that needed to be asked, to Dan Coster for thoughtful proofreading and editorial assistance, to Doug Cooper for many helpful suggestions, and mostly to Yvette Malamud, my editor, who labored long, hard and faithfully and without whom this book would never have made it to press.

To The Student

How to Use this workbook

First, read the Preface. It tells you how the workbook is organized. You should use this workbook chapter by chapter in parallel with the main book. It will help you better understand the material. It will also give you many problems on which you can test your understanding.

After reading a chapter in the main book, turn your attention to the same chapter in the workbook. Work through this workbook chapter. You should have no trouble with the problems immediately following the examples. After you have mastered those, try each of the Exercises. Then re-read the chapter in the main book.

Mathematical Preparations

before you begin work in elementary statistics, you should make certain that you have the necessary quantitative skills. You will not need any calculus or algebra—neither FPP nor the workbook use anything more than simple arithmetic and the most rudimentary word-formulas. What you will need, however, is a good feeling for numbers and for the operations of ordinary arithmetic.

The following diagnostic quiz will help you decide if you are adequately prepared. In each case, circle the choice which *most closely* answers the question. Although you may not be able to determine the exact answer, only one of the multiple choice answers will be reasonable. You should be able to answer all these questions without calculating. Give yourself 20 minutes to answer all the questions.

1. 5/16 is closest to: a) 1/2; b) 1/3; c) 1/4; d) 1/5.

2. The cube of 1/3 is: a) 3/9; b) 3/27; c) 1/81; d) 1/27.

3. Of the following sequences of fractions, what set is arranged in increasing order?

   a) 7/12, 5/6, 2/3, 3/4
   b) 7/12, 7/11, 8/11, 4/5
   c) 7/12, 8/11, 8/12, 9/12
   d) 7/12, 1/2, 7/15, 1/3

4. 13/38 is approximately: a) 5%; b) 15%; c) 25%; d) 35%; e) 45%; f) 55%; g) 65%; h) 75%; i) 85%; j) 95%.

5. When 0.3 is divided by 1/4, the answer is approximately: a) 0.75; b) 0.075; c) 1.2; d) 0.12.

6. Of the following, the value closest to that of $(42.10 \times 0.0003)/0.002$ is: a) 0.063; b) 0.63; c) 6.3; d) 63.0.

7. The number $(1.20672 \times 2.00012)/0.0502698$ is:
   a) nearly but not quite 50
   b) somewhat more than 50
   c) slightly less than 100
   d) more than 100

8. 72.2376% of 416.9327 is approximately: a) 200; b) 300; c) 400; d) 500.

9. A is older than B. With the passage of time, the ratio of the age of A to the age of B: a) remains the same; b) increases; c) decreases.

10. 300 is what percent of 1500? a) 2%; b) 5%; c) 10%; d) 15%; e) 20%; f) 25%; g) 30%; h) 50%; i) 75%

11. 0.125 written as a percent is: a) 1/8%; b) 0.125%; c) 12.5%; d) 125%.

A is the shaded region in the above square.

12. What is the approximate ratio of the area of A to the area of the square?
   a) 0.1; b) 0.3; c) 0.7; d) 0.9.

13. Of the following, which best describes the proportion of A that is in the left half of the square? a) 23/71; b) 1/71; c) 70/71; d) 37/71

A car is driven at constant speed from San Francisco to Los Angeles by way of Palo Alto. Here are four factors of interest to the driver.

14. The distance from San Francisco after one hour

15. The time required to go 100 miles

16. The distance from Palo Alto after one hour

17. The distance from Los Angeles after one hour

Following are six graphs. In each, the horizontal axis represents the speed of the car. In four of the graphs, the vertical axis is one of the factors listed above. In questions 14 through 17, match the factor with the

To The Student

proper graph (a, b, c, d, e, or f).

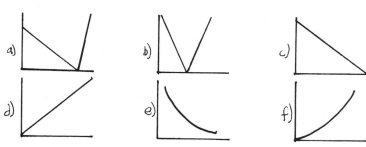

18. Of the following, the pair that is *not* a set of equivalents is:

   a) 0.021%, 0.00021; b) 1/4%, 0.0025; c) 1.5%, 3/200; d) 225%, 0.225

19. One-fourth percent of 360 is: a) 0.09; b) 0.9; c) 9.0; d) 90.

20. The relationships between 0.01% and 0.1 is: a) 1 to 10; b) 1 to 100;
c) 1 to 1,000; d) 1 to 10,000.

21. Vodka is 40% alcohol. If one quart of vodka is mixed with 0.6 quarts
of orange juice, what percentage of the mixture is alcohol?

   a) 15%; b) 20%; c) 34%; d) 25%; e) 6¾%; f) cannot be determined
   from the data given

22. Four quarts of a certain mixture of alcohol and water is at 50%
strength. To it is added a quart of water. The alcohol strength of the
new mixture is: a) 12.5%; b) 20%; c) 25%; d) 40%.

23. The square root of 100,000 is about: a) 30; b) 300; c) 1,000;
d) 3,000; e) cannot tell.

24. The closest approximation to the square root of 2/5 is: a) 0.65;
b) 1/5; c) 0.16; d) 0.4.

Now grade yourself—the answers are: 1b; 2d; 3b; 4d; 5c; 6c; 7a; 8b;
9c; 10e; 11c; 12b; 13d; 14d; 15e; 16b; 17c; 18d; 19b; 20c; 21d;
22d; 23b; 24a.

If you scored less than 16, you need more preparation. I recommend that
you obtain a copy Of <u>Mathematics</u> <u>for</u> <u>Statistics</u> by W. L. Bashaw (John
Wiley & Co.). Chapters 1 through 6, 18, 19, 21 and 22 will give you the
information and practice you need. You should master all of this material
before you begin the main book or the workbook.

To the Student

A Note on the Approach

FPP is unusual in its approach to statistics. It attempts to give you a real understanding of the ideas of the subject. It avoids formulas and focuses on reasoning and intuition. On the face of it, the book looks a lot less imposing than most mathematics textbooks. However, there is a price to pay—in the absence of formulas you must learn to think. In some ways, this book is harder than a book filled with formulas and equations. So don't be deceived: although many pages are mostly pictures and prose, they may require a lot of your attention.

Many of the problems in Chapters 3 through 12 emphasize a geometrical approach to the material. It is important to have a good grip on this—if you do, many relationships that otherwise might puzzle you will become clear. I encourage you not to neglect these problems. Work with your pencil and draw all the pictures suggested in the workbook—the reward will be a solid understanding which will carry you beyond memorized procedures.

1

Controlled Experiments

I. Exercises

1.1 An experimental program to improve reading skills in school children is set up with an aftertest to determine its effect. The children are to be divided into two groups: one to receive instruction in the program and the other to receive only their ordinary reading instruction. Describe a method for deciding how to allocate the children to the "treatment" group (the experimental program) and the control group (with ordinary reading instruction) which will not introduce any bias.

1.2 It is conjectured that a food supplement improves the intelligence of rats as measured by their ability to run a maze. From a cage containing twenty test rats, ten are chosen to belong to the control group and ten are chosen to belong to the treatment group. The ten for the control group are chosen by the lab assistant who pulls out the ten rats nearest the cage door. What kind of bias might be introduced by this selection process?

II. Solutions

1.1 The children should be assigned to treatment and control groups at random. You could for instance put all their names in a box and draw out half the names half the names at random—then assign them to the treatment group and those whose names weren't drawn to the control group.

2

Observational Studies

I. Exercises

2.1 An advertisement states that according to a study conducted by an automobile association 90% of all Volvo trucks sold in the last ten years are still on the road. Does this necessarily mean that Volvo trucks last a long time or could there be another explanation?

2.2 In 1979 surveys showed that in Alcoholics Anonymous there are now half as many members under the age of 30 as there were three years ago. This may be due to a decrease in alcoholism among those under 30. Can you think of another contributing factor?

2.3 An English survey of 3000 medical records showed that smokers are more inclined to get depressed than non-smokers. Does this necessarily imply that smoking causes depression?

2.4 Hospital records show that twice as many men as women fall out of hospital beds. Does this indicate that men are clumsier than women or is there another possible explanation?

2.5 Studies show that in the period from 1850 to 1900 in the United States the average marriage lasted only 12 years. Does this show that the divorce rate was high in that period?

2.6 In an investigation of the incidence of heart disease in men it is noticed that the fathers of one-child families have had on the average somewhat fewer heart attacks than the fathers of five-child families. It is concluded that having big families increases one's chance of attack. Do these facts support the conclusion?

2.7 The usual method of catching tuna with nets accidentally results in the capture (and death) of dolphins. In response to reports that the dolphin is in danger of extinction a new kind of net for catching tuna is introduced. It is tried for one year and the percentage of dolphin out of the total tonnage caught is compared with the average percentage caught over the preceding ten-year period using the old method. A large decrease in the percentage of captured dolphins is observed. This is for-

Solutions

warded as evidence that the new method helps to protect dolphins. Suggest an alternate interpretation.

2.8 Columnist Ann Landers was asked whether having children was worth the problems involved. She asked her readers, "If you had to do it over again would you have children"? A few weeks later her column was headlined, "*Seventy Percent of Parents Say Kids Not Worth It*", because 70% of the parents who wrote said they would not have children if they could make make the choice again. Is this a valid conclusion? Comment.

II. Solutions

2.1 It could be that most trucks were sold this year or in recent years. Or Volvo drivers could use their trucks very little. There are many other possible explanations.

2.3 No—it may be that people with a tendency towards depression also tend to smoke.

2.5 No—the shortened marriages were mostly due to the death of one of the partners.

2.7 It may be that the number of dolphins in the sea has already been reduced by the previous years' catches.

3

The Histogram

1. Reading Histograms

To learn how much of a data set lies between two values, shade in the area above these values on the histogram and estimate the proportion of the histogram that is shaded.

Example A: Below is a rough histogram of the test scores in a class.

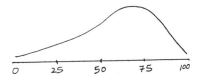

What percentage scored above 75?
Solution: Shading in the area above 75 gives the following picture:

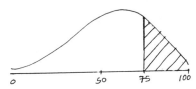

The shaded region appears to be about 30% of the histogram, area-wise. So the answer is: about 30%.

Problems

3.1 Estimate the percentage scoring below 25.

3.2 Estimate the percentage scoring between 50 and 75.

Reading Histograms

Example B: Below is a histogram of heights of all family members in a study of families in which the parents have been married less than five years. How do you explain the two peaks?

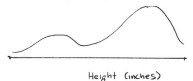

Height (inches)

Solution: The large peak to the right is the parents' heights; the small peak to the left is the heights of their children, who are less in number.

Example C: Below is a histogram of the weights of all people in a grammar school. Explain the large left peak. and the smaller right peak.

Solution: The larger left peak comes from the weights of the children; the smaller right peak comes from the weights of the (fewer) teachers.

3.3 Here are three data sets:

I. Pounds of beef consumed per household in USA; II. Pounds of beef consumed per household in India; III. Pounds of beef consumed per household in a city (in the USA) having a small wealthy population and a large poor population.

Here are the histograms of these data sets in scrambled order:

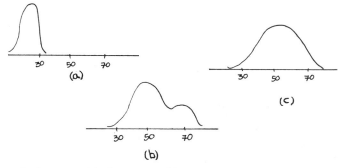

Match each data set with the correct histogram.

5

3.4 Here are six histograms:

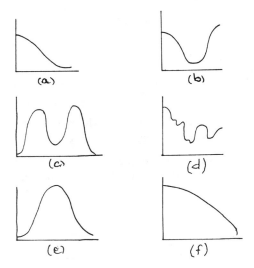

They come from a large random sample of adults living in California. For each person in the sample, four numbers are recorded: I. The person's height; II. The person's weight; III. The distance from the person's home to San Francisco (to nearest mile); IV. The distance from the person's home to the nearest airport.

Match each data set with the correct histogram. You may use each histogram more than once or not at all.

3.5 The following are two histogram sketches representing daily air pollution levels over Los Angeles, *before* and *after* legislation was enacted to curb noxious exhaust emissions from motor vehicles.

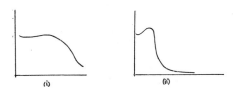

a) Identify on the sketch which curve is which (assume the law was effective); b) Suppose the *before* and *after* data were combined into one list; is it possible to sketch the resulting histogram? Do so, or explain what information is lacking.

6

Drawing Histograms

II. Drawing Histograms

Usually you are given class intervals and the percentages of the data falling in each interval. The trick here is to adjust the percentages found in the various intervals. This is done by choosing a "common unit" for the data, and expressing the density in each interval as a percentage per common unit. You may choose any common unit you like, but the arithmetic comes out more simply if you choose a common unit which divides easily into each of the interval lengths. As an example consider the following problem:

Example A: A distribution table of GPA's for a student group is shown below. Plot the histogram for this table. Mark the horizontal and vertical scales carefully.

GPA	%
0.0 - 1.0	10
1.0 - 2.0	20
2.0 - 3.0	40
3.0 - 4.0	30

1) What is a natural common unit? *1.0 GPA point*

2) List the number of common units in each interval next to the "%" column and label this column "number of common units":

GPA	%	number of common units	
0.0 - 1.0	10	1	(since 0.0 - 1.0 is 1 GPA point)
1.0 - 2.0	20	1	(since 1.0 - 2.0 is 1 GPA point)
2.0 - 3.0	40	1	(since 2.0 - 3.0 is 1 GPA point)
3.0 - 4.0	30	1	(since 3.0 - 4.0 is 1 GPA point)

3) Divide the "%" column by the "number of common units" column and call this the "percent per common units" column:

GPA	%	number of common units	percent per common units
0.0 - 1.0	10	1	10
1.0 - 2.0	20	1	20
2.0 - 3.0	40	1	40
3.0 - 4.0	30	1	30

4) Draw a horizontal scale with all intervals marked.

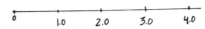

5) Draw a vertical scale which will accommodate all values in the "percent per common units" column:

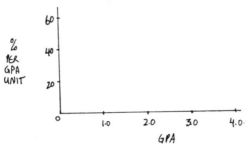

6) Draw boxes over each interval with height equal to that interval's "percent per common units":

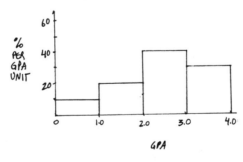

Example B: A distribution table of GPA's for a student group is shown below. Plot the histogram for this table. Mark the horizontal and vertical scales carefully.

GPA	%
0.0 - 2.0	10
2.0 - 2.5	20
2.5 - 3.5	40
3.5 - 4.0	30

1) What is a natural common unit? *1.0 GPA point*

2) List the number of common units in each interval next to the "%" column and label this column "number of common units":

Drawing Histograms

GPA	%	number of common units	
0.0 - 2.0	10	2.0	(since 0.0 - 2.0 is 2.0 GPA points)
2.0 - 2.5	20	0.5	(since 2.0 - 2.5 is 0.5 GPA points)
2.5 - 3.5	40	1.0	(since 2.5 - 3.5 is 1.0 GPA points)
3.5 - 4.0	30	0.5	(since 3.5 - 4.0 is 0.5 GPA points)

3) Divide the "%" column by the "number of common units" column and call this the "percent per common units" column:

GPA	%	number of common units	percent per common units
0.0 - 2.0	10	2.0	5 (= 10/2)
2.0 - 2.5	20	0.5	40 (= 20/0.5)
2.5 - 3.5	40	1.0	40 (= 40/1)
3.5 - 4.0	30	0.5	60 (= 30/0.5)

4) Draw a horizontal scale with all intervals marked.

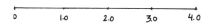

5) Draw a vertical scale which will accommodate all values in the "percent per common units" column:

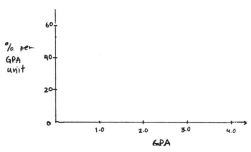

6) Draw boxes over each interval with height equal to that interval's percent per common units:

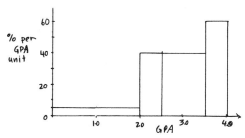

Note: The choice of a common unit is up to you—choose one that is convenient. Whichever one you choose, the histogram will look the same, except for the scale. For instance, if you had chosen 0.5 as the common unit, the table in step 3) would look like this:

GPA	%	number of common units	percent per common units
0.0 - 2.0	10	4	2.5
2.0 - 2.5	20	1	20
2.5 - 3.5	40	2	20
3.5 - 4.0	30	1	30

The histogram would look like this:

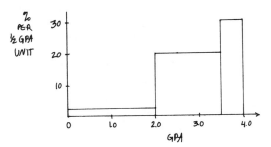

Problems

3.6 A distribution table of monthly wages for part-time employees is shown below. No one earned more than $1,000 a month. The class intervals include the left endpoint, but not the right. For example, the second line of the table says that 20% of the wages were $100 or more, but less than $200.

Dollars	Monthly
0.0 - 100	10
100 - 200	20
200 - 400	40
400 - 1000	0

1) What is a natural common unit? _____

2) List the number of common units in each interval next to the "%" column and label this column "number of common units":

Dollars Monthly	%	number of common units

Drawing Histograms

3) Divide the "%" column by the "number of common units" column and call this the "percent per common units" column:

Dollars	%	number of common units	percent per common units

4) Draw a horizontal scale with all intervals marked.

5) Draw a vertical scale which will accommodate all values in the "percent per common units" column:

6) Draw boxes over each interval with height equal to that interval's "percent per common unit":

3.7 The distribution of the number of cars per family in a certain area is:

number of Cars	number of Families
0	8
1	22
2	15
3	5

Note that in this problem you are given the actual number of data points in each class interval—you must first convert them to percentages.

1) What is a natural common unit? _____

2) List the number of common units in each interval next to the "%" column and label this column "number common units":

number of Cars	%	number of common units

3) Divide the "%" column by the "number of common units" column and call this the "% per common units" column:

number of Cars	%	number of common units	percent per common unit

4) Draw a horizontal scale with all intervals marked.

Drawing Histograms

5) Draw a vertical scale which will accommodate all values in the "percent per common unit" column:

6) Draw boxes over each interval with height equal to that interval's "percent per common unit":

3.8 A distribution table for the scores on an exam is shown below. The classes include the left endpoint, but not the right. For example, the second line of the table says that 20% of the people scored 50 points or better, but less than 70 points.

Points	%
0 - 50	10
50 - 70	20
70 - 80	25
80 - 90	30
90 - 100	15

Draw a histogram of this data.

III. Exercises

3.9 True/False

An example of a qualitative measurement is the amount of butterfat in milk.

3.10 Which of the following are qualitative measurements? Which are quantitative?

a) shoe size; b) hair color; c) sex; d) smoker or non-smoker; e) number of cigarettes smoked per day; f) air temperature; g) make of car; h) miles per gallon; i) price of gasoline; j) IQ

3.11 Which of the following quantitative variables are discrete and which are continuous?

a) shoe size; b) air temperature; c) IQ; d) GPA; e) years of schooling completed; f) rainfall in inches; g) number of TV's owned

3.12 Here is a histogram of test scores.

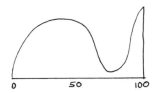

True/False

a) More than half the scores were over 50; b) There were more scores in the range 90-100 than in the range 70-80.

3.13 The histogram for educational level (years of schooling) of a representative cross-section of Americans looks like this:

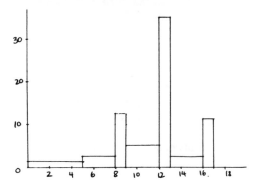

a) The percentage that finished high school and didn't go on is closest to: i) 70%; ii) 35%; iii) 15%; iv) 10%.

b) The percentage having more than a high school education is closest to: i) 80%; ii) 40%; iii) 20%; iv) 10%.

c) How do you explain the peaks over 8, 12 and 16 years?

3.14 As part of a study on the effects of flu, some doctors took the temperatures of 200 flu patients in their third day of illness. They obtained the following histogram:

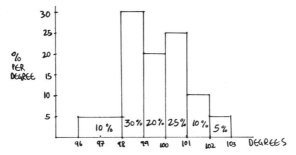

a) From another part of the hospital, another 50 patients are found to have temperatures between 96 and 98. These patients are combined with the rest and a new histogram is drawn for all 250 patients. The height of the rectangle with base 96-98 degrees becomes: i) 24% per degree; ii) 28% per degree; iii) 14% per degree; iv) 12% per degree; v) can't tell without knowing for the new patients how many had temperatures between 96 and 97 degrees.

b) The number of patients who had temperatures below normal (98.6 degrees) was approximately: i) 28; ii) 92; iii) 56.

3.15 As part of a study on the effects of a weight-losing program, the weights of 200 people in the program were recorded upon entry. The following histogram was obtained:

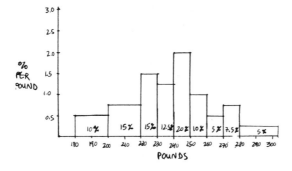

a) Two days later, another 50 people enter the program and all are found to have weights between 200 and 220 lbs. These people are combined with the rest and a new histogram for entry weights is drawn for all 250 people. The height of the rectangle with base 200-220 becomes: i) 0.5% per lb; ii) 1.2% per lb; iii) 1.6% per lb; iv) 2.8% per lb; v) can't tell without knowing for the new people how many had weights between 200 and 220 lbs.

b) In the original histogram, the number of people who had weights below 210 was approximately: i) 28; ii) 35; iii) 69.

c) What percent of the original 200 people had weights between 200 and 220 lbs?

d) How many of the original 200 people had weights between 200 and 220 lbs?

e) How many of the 250 people had weights between 200 and 220 lbs?

f) What percent of the 250 people had weights between 200 and 220 lbs?

g) What is the height of the rectangle with base 200-220 in the new histogram?

3.16 Part of a histogram for family income in a certain city is shown below (density shown in parentheses).

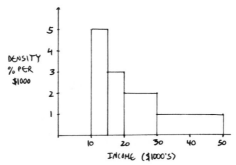

a) What percent of the families earned between $15,000 and $30,000?

b) What percent of the families earned between $15,000 and $35,000?

Exercises

3.17 Below is a histogram of monthly wages for part-time employees. No one earned more than $1,000 (the data is hypothetical). The block over the class interval from $200 to $500 is missing. How tall must it be?

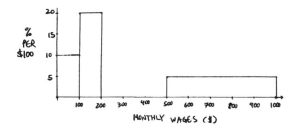

3.18 Here are two histograms. One of them represents the age of death by suicide and the other represents the age of death by any means, suicide included. Which is which?

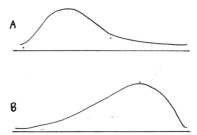

3.19 A distribution table for the scores on an exam are shown below. The class intervals include the left endpoint, but not the right. For example, the second line of the table says that 20% of the people scored 50 points or better, but less than 70 points.

Points	%
0 - 50	10
50 - 70	20
70 - 80	25
80 - 90	30
90 - 100	15

Sketch the histogram over the first two class intervals (0-50, 50-70).

3.20 A distribution table for the age of passenger cars in the US in 1975 is shown below. The class intervals include the left endpoint, but not the right. For example, the second line of the table says that 35 percent of the cars were 5 years old, but less than 10 years old.

Age (years)	%
0 - 5	45
5 - 10	35
10 - 20	20

Plot the histogram for this table. Mark the horizontal and vertical scales carefully.

3.21 The subjects in a certain study range in age from 21 to 60 years. The first two rows of the age distribution are shown below:

Age (years)	%
21 - 22	10
23 - 34	12

The interval 21-22 includes the subjects 21 or older, but not yet 23. The second interval is interpreted similarly. Sketch the histogram over these two class intervals. Mark the numerical scales on the horizontal and vertical axes completely.

IV. Solutions

3.1 The picture looks like:

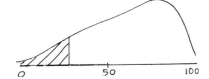

A reasonable estimate is: about 15%.

3.2 The picture looks like:

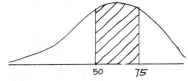

A reasonable estimate is 35% or 40%.

Solutions

3.3 II is a, since Indians eat very little beef. III will lie to the left of I, since poor people cannot afford as much beef as the general population—so III is b and I is c.

3.4 I—e, II—e, III—d, IV—a

3.5 a) (i) is before, (ii) is after. b) No—the number of observations in each case would be needed. The "before" histogram might contain 90% of the data, in which case the new histogram would look almost identical—or it could contain only 10% of the data, in which case the new histogram would look more like the "after" histogram.

3.6 1) $100 is the most natural common unit.

2) and 3)

Dollars	%	number of common units	percent per common units
0 - 100	10	1	10
100 - 200	20	1	20
200 - 400	40	2	20
400 - 1000	30	6	5

4), 5) and 6):

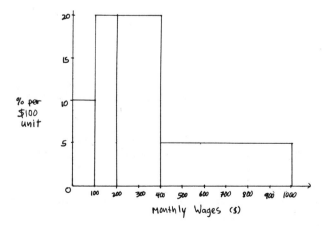

19

3.7 1) 1 car

2) and 3):

number of Cars	%	number of common units	percent per 1 car unit
0	16	1	16
1	44	1	44
2	30	1	30
3	10	1	10

4), 5) and 6):

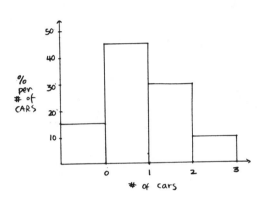

3.8

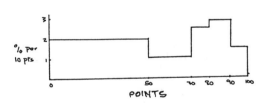

3.9 False—this is a quantitative measurement.

3.11 Shoe size, years of schooling, and number of TV's owned are discrete— all the others are continuous.

3.13 a) Those who finished high school and didn't go on are in the box over 12—35%; b) Those having more than a high school education are in the boxes over 13, 14, 15 and 16—about $3 \times 3\% + 12\% = 20\%$; c) 8 years of schooling represents grammar school, 12 represents high school completion, and 16 represents college completion. These are the most usual times to end education.

Solutions

3.15 a) iii—1.6% per lb; b) ii—35; c) 15%; d) 30; e) 80; f) 32%;
g) 1.6% per pound

3.17 The height is 15% per $100. The blocks account for 10% + 20% +
25% = 55% of the group. This leaves 45% spread over three $100 inter-
vals, so the height over these intervals is 15% per $100.

3.19 Using 10 points as the common unit:

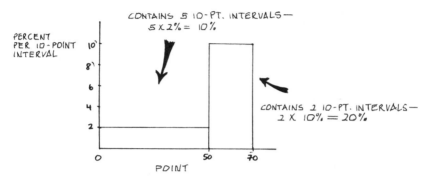

3.21

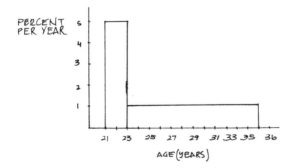

21

4

The Average and Standard Deviation

I. Computing the Average

Here is a program for computing the average of a list.

1) Add up all the numbers in the list.
2) Divide by the number of numbers in the list.

Example A: Find the average of: 1, 1, 4.

Solution: 1) $1 + 1 + 4 = 6$; 2) $\frac{6}{3} = 2$

The average is 2.

Problems

4.1 Find the average of: 1, 1, 1, 1, 2, 2, 2, 3, 3, 4.

4.2 Add 3 to each number in the list in the previous problem, obtaining: 4, 4, 4, 4, 5, 5, 5, 6, 6, 7. What is the average of this new list?

4.3 Multiply each number in the list in problem 4.1 by 7, obtaining the new list: "7, 7, 7, 7, 14, 14, 14, 21, 21, 28". What is the average of this new list?

The previous two problems illustrate two simple facts—namely, if you add the same value to all numbers in a list, the average of the list is increased by this same value; if you multiply all numbers in a list by the same value, the new average is just the old average multiplied by that value. The same holds for subtraction and division—whatever operation which you perform on the numbers in the list is performed on the old average to get the new average. If you perform two operations in succession on the numbers in a list, the same operations in the same succession change the old average into the new average.

Finding the Sum from the Average

Example B: The average of the list "1, 4, 4" is 3. Take each number in this list and first *multiply it by 5, then add 2*, obtaining the new list "7, 22, 22". What is the new average?

Solution: The rule above says that you perform the same operation on the old average: *multiply it by 5, then add 2.* This gives $5 \times 3 + 2 = 17$; the average of the new list "7, 22, 22" is 17. You can check this by:
$$\frac{7 + 22 + 22}{3} = \frac{51}{3} = 17.$$

Example C: There are 2.54 centimeters per inch. A group of 10 people has the following heights (to the nearest inch): 68, 70, 67, 71, 68, 67, 68, 70, 67, 67. The average of these heights is 68 inches. Suppose these heights are converted to centimeters. What is the average height in centimeters?

Solution: The list of heights in centimeters is obtained by multiplying each number in the original list by 2.54: "172.72, 177.80, 170.18, . . . " You could, of course, get the answer by averaging these new numbers. However, it is shorter and simpler to use the rule we have been discussing—multiply the old average by 2.54. The answer is thus: $68 \times 2.54 = 172.72$.

Problem

4.4 To go from temperature in Fahrenheit to temperature in Centigrade subtract 32 degrees from the Fahrenheit temperature and then multiply by 5/9. Thus, to convert 212 degrees Fahrenheit (the boiling point of water) to Centigrade, compute (212 − 32 degrees Fahrenheit) × 5/9 = (180 degrees Fahrenheit) × 5/9 = 100 degrees Centigrade. The average of a list of daily temperatures in Tucson is 77 degrees Fahrenheit. What is the average in Centigrade?

II. Finding the Sum From the Average

The following principle is often used in "tricky" average problems. To find the sum of the numbers in a list, multiply the average by the number of numbers in the list.

Example A: The average of the list "3, 3, 3, 3, 4, 4, 4, 5, 5, 6" is 4. What is the sum?

Solution: There are 10 numbers in the list, so the sum must be $10 \times 4 = 40$.

Example B: A list of twenty numbers has average 100. One of the numbers is increased by 10. What is the new average?

Solution: To compute the new average, you would first compute the new sum, and then divide by 20. What is the new sum? Well, it must be the old sum plus 10 since only one number was increased by 10. The old average was 100, so the old sum was $20 \times 100 = 2000$. Thus the new sum is $2000 + 10 = 2010$, and the new average is $2010/20 = 100.5$.

Problems

4.5 The average of your ten quiz scores in a course is 80. A mistake in grading is found on one quiz and you receive 10 more points on that quiz. What is your new average quiz score?

4.6 A swim team has five members. Their average time in the 100-yard free-style is 55 seconds. One of the members is dropped. Her time was 63 seconds. What is the average time of the remaining four members?

III. Locating the Average on a Histogram

To locate the average on a histogram, imagine the histogram is made of wood. Move your finger from left to right along the base of the histogram until you feel that the blocks of wood will balance on your finger—this balance point is the average.

Example A: Estimate the average of the following histogram.

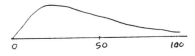

Solution: The histogram appears to balance with your finger placed a little to the left of 50—so a reasonable estimate of the average is just that: a little less than 50. Forty or 45 would be acceptable guesses.

Example B: On the histogram below, three points are marked. One of them is the average—which one?

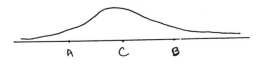

Computing the Median

Solution: If you place your finger at A, the histogram will tip to the right. If you place it at B it will tip to the left. Only at C does it appear to balance—C is the average.

Problem

4.7 Which of the following two histograms has a larger average?

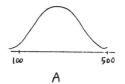

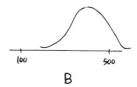

A   B

IV. Computing the Median

Computing the median is easy but tedious. The median is the "middle value" in a list—it is the halfway point from lowest value to highest value. To compute the median:

1) Rewrite the list in increasing order.
2) Find the middle number in the list.

Example A: Find the median of the list: 13, 23, 7, 8, 2.
Solution: Rearranging the list in increasing order, we get: 2, 7, 8, 13, 23. The middle number is 8, which is the median.

Example B: Find the median of: -10, 10, 20, -3, 4, 7, 15.
Solution: The list in increasing order is: -10, -3, 4, 7, 10, 15, 20. The middle number—and hence the median—is 7.

Example C: Find the median of: 13, 23, 7, 8.
Solution: The list in increasing order is 7, 8, 13, 23. There is no "middle number"—this will happen whenever the list has an even number of numbers. So what is the median? In this case, statisticians call any number *between* 8 and 13 a median. For instance, 9 is a median; so is 10.6; so is 12. Generally, we will be dealing with larger lists, and the median(s) will be either unique or so close together relative to the spread of the data that we will speak of one "median"—the "middle" of the histogram.

Problem

4.8 Find the median of: 130, 2, 18, 38, 7.

V. Locating the Median on a Histogram

To locate the median on a histogram, simply move your finger from left to right under the histogram until 50% of the histogram area is to the left of your finger, and 50% is to the right.

Example A: Find the median of the following histograms:

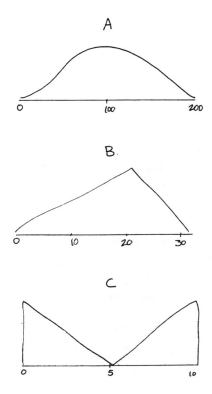

Solution: A has median 100; B's median is about 15 or so; C's is 5.

Locating the Median on a Histogram

Example B: Here are two histograms. In each case, say whether the median is less than 10, equal to 10, or greater than 10.

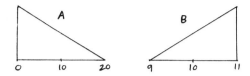

Solution: If you place your finger at 10 under A, more than half the area is to the left—so you have moved your finger too far from left to right. The median is less than 10. In B, your finger at 10 has not moved far enough to the right—it still has less than half the area to it's left. So the median is greater than 10.

Example C: Here is a histogram of the weights of 3581 women in a Health Examination Survey. Approximately what is the median weight?

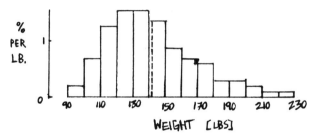

WEIGHT [LBS]

Solution: The median appears to be a little less than 140 lbs.

Problem

4.9 Estimate the median of the following histogram:

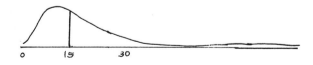

27

VI. The Average and the Median

The average and the median are both measures of location—the center of the histogram. You should gain some insight into their relationship, and into what they tell you about the histogram. The average can be either larger or smaller than the median—there is no rule. If the average is either much larger or much smaller than the median, the histogram is skewed, with one tail very long. For instance, the following histogram will not balance at the median—if a fulcrum were placed there, it would tip to the right. So the average is larger than the median—the histogram's long right-hand tail pulls the average off to the right.

Example A: Here are two histograms. In which case, A or B, is the average larger? In which case is the median larger?

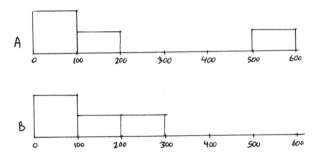

Solution: Histogram A has the larger average; but the median in each case is the same—100.

Example B: How could you change histogram A (above) so that the median would remain the same but the average would increase? How large could you make the average without changing the median?

Solution: Sliding the right-most block further to the right increases the average, but does not change the median. There is no limit to how large you can make the average without changing the median. For instance, the following histogram increases the average to almost 250,000:

Computing the RMS

Example C: Would you expect average U.S. income or median U.S. income to be larger?

Solution: Since the income histogram has a long right-hand tail due to a few people with extremely high incomes, average income should be larger.

Problems

4.10 In a data set the median is much larger than the average. The histogram for this data set has a:

a) long right-hand tail; b) long left-hand tail; c) symmetrical shape.

4.11 In a data set the median is much smaller than the average. The histogram for this data set has a:

a) long right-hand tail; b) long left-hand tail; c) symmetrical shape.

VII. Computing the RMS

The following program will help you compute the rms of a list.

1) Square each number in the list.

2) Average the squares.

3) Take the square root.

Example A: Find the rms of the list: 1, 4, 5.
Solution: 1) 1, 16, 25 are the squares

2) $\dfrac{1+16+25}{3} = 14$ is the average of the squares

3) rms $= \sqrt{14}$

Example B: Find the rms of the list: $-1, -4, +5$.
Solution: 1) 1, 16, 25 are the squares

2) $\dfrac{1+16+25}{3} = 14$ is the average of the squares

3) rms $= \sqrt{14}$

Notice that the lists in Examples A and B are identical if you ignore the signs—the rms in B is the same as in A because rms measures size without regard for sign.

Problems

4.12 Find the rms of the list: $-1, -1, -1, -1, 4$.

4.13 Find the rms of the list: $-3, -1, -1, 5$.

VIII. Estimating the RMS

You should be able to look at a list of numbers and make a rough guess of its rms. You can think of the rms as the "typical size" of numbers in the list ignoring sign.

Example A: The rms of the list: "1, 4, 5" is closest to:

a) 1; b) 3.5; c) 5; d) 0.

Solution: The answer closest to a typical number from this list is 3.5. The actual rms is the $\sqrt{14}$, as we saw above, which is equal to 3.74.

Problems

4.14 First estimate, then compute the rms of each of the following lists:

a) $+3, -5, -4, +6, +7$

b) $-10, -20, +15, -9, +18, -12$

c) $-100, +100, -90, +110, +120$

4.15 Here are four lists and their four rms values in jumbled order. Match each list with its rms, without calculation.

a) $-4, +3, +2, +4, -3, -2$

b) $+10, +8, +5, +12, +17, +10$

c) $-5, -5, +5, -5, +5$

d) $+15, +16, -16, -14, -15$

The rms values are: 5, 15.3, 3.1, 11.0.

IX. Computing the SD

The following program will help you compute the SD of a list.

1) Find the average of the list.

2) Calculate the deviations from the average by subtracting the average from each number in their list.

3) Square the deviations.

4) Find the average of these squares.

5) Take the square root of step 4)—this is the SD.

Estimating the SD from a List

Example A: Calculate the SD of the list: 1, 5, 5, 7, 7, 11.
Solution: 1) Average = 6

    2) Deviations from the average are: $-5, -1, -1, 1, 1, 5$

    3) Deviations squared are: 25, 1, 1, 1, 1, 25

    4) Average of the squares = 9

    5) Square root of this average = 3, so the SD = 3.

Example B: Find SD of list: $-2, -1, -1, 0, -1, 0, +1, -2, -2, -2$.
Solution: 1) Average = -1

    2) Deviations from average: $-1, 0, 0, 1, 0, 1, 2, -1, -1, -1$

    3) Deviations squared: 1, 0, 0, 1, 0, 1, 4, 1, 1, 1

    4) Average of squares = 1

    5) Square root of average = 1, so the SD = 1.

Problems

4.16 For each list below, work out the average, all the deviations from the average, and the SD.

    a)   2, 6, 8, 10, 14

    b)   4, 10, 18, 16, 22

4.17 Find the average and SD in each of the following lists:

    a) $-1, -1, -1, -1, 0, 0, 0, 1, 1, 2$

    b) $-1, -1, -1, -1, 4$

    c) $-3, -1, 0, 1, 3$

    d) $-3, -1, -1, 5$

    e) $-5, -1, -1, 1, 1, 5$

    f) $-5, -4, -2, 0, 0, 0, 0, 2, 4, 5$

    g) $-4, -4, -3, -2, 0, 0, 2, 3, 4, 4$

    h) $-11, -10, -9, -3, -3, 3, 3, 9, 10, 11$

X. Estimating the SD from a List

It is important to be able to estimate the SD of a list. You do this by first estimating the average and then looking for "typical" deviations from the average.

Example A: Here is a list of numbers: 4, 5, 5, 5, 4, 6, 7, 6, 4, 4. What is the SD, roughly?

Solution: The average of the numbers is about 5; they deviate from 5 by 1 or so. A rough estimate of the SD is therefore 1. Calculation shows that the SD is precisely 1—your guess won't always be this accurate.

Problem

4.18 Here are four lists and their four SD's in jumbled order. Match each list with its SD.

a) 0, 4, 4, 6, 6, 10

b) 0, 1, 2, 8, 8, 14, 14, 20, 21, 22

c) 0, 0, 0, 0, 1, 1, 1, 2, 2, 3

d) −20, −20, −20, −20, 0, 0, 0, 20, 20, 40

SD's: i) 1; ii) 3; iii) 8; iv) 20

XI. Estimating the SD from a Histogram

It is also important to have a geometrical picture of the SD. If a histogram is symmetrical, you can estimate the SD roughly by locating the average (which is at the exact center if the histogram is truly symmetrical) and then moving your eye (or finger) out from the center until you reach a point with about half the area further from the average and half the area closer to the average. This gives a crude—but fairly effective—estimate of the SD.

Example A: Give a rough estimate of the SD of the following histogram.

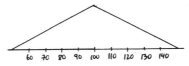

Solution: The average is 100. Moving down to a little less than 90, say to about 87,

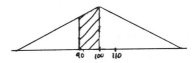

it appears that the shaded region represents about half the area to the left of the average, and hence about half the deviations are smaller (i.e., in the non-shaded region labeled A). So a rough estimate of the SD is: 100 − 87 = 13.

Exercises

Problems

4.19 Here are three histograms and three line segments representing their SD's. Their order is jumbled. Match each histogram with the line segment representing its SD.

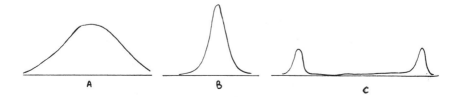

i)   ————————

ii) ————

iii) —

4.20 Which of the following two histograms has a larger standard deviation?

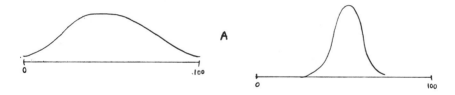

XII. Exercises

4.21 True/False

a) The average height in a class is 60 inches. There are 30 people in the class. One can conclude that 15 of the students are less than 60 inches tall.

b) The average of a list of numbers cannot be smaller than its SD.

c) If two lists have the same average and SD, their histograms must be the same.

d) The SD is never negative.

e) If the SD of a list of numbers is zero, then its average must also be zero.

f) If the SD of a list of numbers is zero, then the numbers must all be the same.

g) If ten is added to each number in a list, the SD is increased by ten.

h) If the right-hand tail of a histogram is longer than the left-hand tail then the median is greater than the average.

4.22 If the rms of a list is zero, which of the following are correct?

a) All the numbers in the list are the same; b) The average is also zero; c) The SD is also zero; d) The median is also zero.

4.23 In which case(s) must the rms of a list be equal to the SD of the list?

a) The SD is zero; b) The average equals the median; c) The average is zero; d) The rms is zero.

4.24 If the SD of a list is zero, which of the following are correct?

a) The average is also zero; b) All the numbers in the list are zero; c) All the numbers in the list are the same; d) The median and the average are the same.

4.25 A list of five numbers has average 6. The first four deviations from the average are 3, 2, −4, and 0. The last number on the list is: a) 0; b) 1; c) 5; d) 2; e) can't tell.

4.26 A class of 400 students is divided into two sections of 200 each. Both sections are given a common final exam. The following is observed:

	Average	SD
Section 1	85	10
Section 2	45	10

Suppose that all 400 scores are combined as one list. The SD of the new list will be: a) smaller than 10; b) larger than 10; c) equal to 10; d) cannot tell anything at all without knowing the entire list.

4.27 A study was made of the age at entrance of college freshmen. The SD turned out to be one of the following: 1 month, 1 year, 5 years. Which one was it? Why?

Exercises

4.28 Here is a histogram with its average marked:

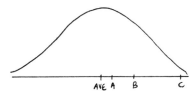

The three points A, B, and C lie at different distances from the average. Which distance best represents one SD?

4.29 Given this group of numbers: 3, 7, 4, 6, 2, 2, 7, 3, 2. Which of the following are the same?

a) average; b) range; c) SD; d) median; e) mode.

Note: the mode is the number appearing in the list the most times, and the range is obtained by subtracting the smallest number in the list from the largest.

4.30 Suppose you are given 10 measurements: 6, 6, 6, 6, 7, 7, 7, 8, 8, 9.

a) For the above numbers, which is the smallest? i) the average; ii) the mode; iii) the median; iv) the range.

b) The SD for the above data is: i) 54/8; ii) 24/8; iii) 20/8; iv) 1.

c) Suppose you discovered that the above data were incorrect. In particular the largest observation was wrong: instead of 9, it should be 18. Which of the following are not affected? i) the average; ii) the median; iii) the SD; iv) the range.

4.31 Suppose a list consists of 3, 3, 3, 4, 4, 5, 7, 8, 8. Two of the following quantities are equal. Which two?

a) average; b) mode; c) median; d) range; e) SD.

4.32 There are 25 non-identical numbers on a list. The average is 20. One number, which happens to equal 20, is removed from the list.

a) The average of the remaining 24 numbers is: i) less than 20; ii) equal to 20; iii) larger than 20; iv) depends on the SD; v) can't tell without knowing all the numbers.

b) The SD of the remaining 24 numbers is: i) less than; ii) greater than; iii) equal to; iv) can't tell without all the data; the SD of the original list.

4.33 Below is a histogram of blood pressures of subjects in a study. The table gives the density over each interval. Average blood pressure for the group is 120mm. and the SD is 10mm. The percentage with blood pressure between 110mm. and 130mm. is: a) 50%; b) 57%; c) 68%; d) 95%.

Interval	Density (%/mm.)
90 − 100	0.4
100 − 105	1.0
105 − 110	2.0
110 − 115	3.0
115 − 120	3.4
120 − 125	2.6
125 − 130	2.4
130 − 135	2.0
135 − 140	1.0
140 − 150	0.6
150 − 160	0.2

4.34 The median of a list is 2.5. If you add ten to each entry in the list what will be the median of the new list?

4.35 Here is a list of numbers: 77, 30, 26, 38, 58, 110, 40.

a) What is the median of this list?

b) Without calculating the average, explain why you expect it to be larger than the median.

4.36 Add a constant to each number on a list. Will the average change? Will the SD change?

4.37 a) Find the SD of the following list: 0, 2, 3, 4, 6.

b) Multiply each value in the above list by 3, obtaining a new list. What is the SD of this new list?

4.38 a) Suppose you increased each number in a list by 5. What would this do to the SD?

b) Suppose you increased each number in a list by 50%. What would this do to the SD?

4.39 The average temperature in Berkeley on the 11th day of May over the last 10 years was 74 degrees F with an SD of 6 degrees F. Temperature Centigrade is found by subtracting 32 degrees from temperature Fahrenheit, then multiplying by 5/9. What were the average and the SD of the temperature in Berkeley on the 11th day of May expressed in degrees Centigrade?

36

Exercises

4.40 a) Make up a list of ten numbers with SD equal to zero.

b) Make up a list of ten numbers with SD equal to one.

4.41 A list contains numbers between 0 and 100. The median of this list is much smaller than the average. Draw a rough sketch of its histogram.

4.42 Estimate the average, median, and standard deviation for a list whose histogram is:

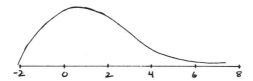

4.43 For the roughly sketched histogram below:

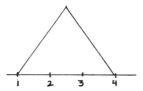

a) What is the approximate average?

b) The SD is about: i) 0.5; ii) 1.0; iii) 2.0.

4.44 Below is a table of the distribution of families by income in the U.S. in 1973. Intervals include the left endpoint, but not the right endpoint.

Income Level ($)	Percent
$ — 1,000	1
1,000 — 2,000	2
2,000 — 3,000	3
3,000 — 4,000	4
4,000 — 7,000	15
7,000 — 10,000	15
10,000 — 15,000	26
15,000 — 25,000	26
25,000 — 50,000	8
50,000 and over	negligible

a) Plot the histogram for the table. Mark the horizontal and vertical scales carefully.

b) What value of the histogram divides the area into halves? c) Does the histogram have a long right-hand tail, a long left-hand tail, or is it symmetrical? d) Without calculation, approximately what is the average income level? e) What is the median? f) Is the average greater than, less than, or about the same as the median? Why? g) Are average and SD a good way to summarize this histogram? If so, why? If not, why not? What do you think might be added? h) The SD for income level is closest to: i) $1,000; ii) $10,000; iii) $15,000.

4.45 A histogram for family income in a city is shown below:

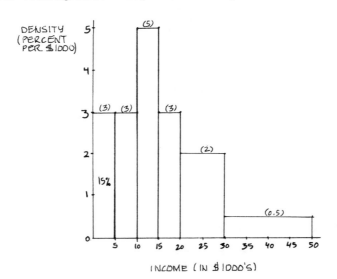

INCOME (IN $1000's)

a) From this histogram, estimate the percentage of people whose income is between $5,000 and $15,000. b) Suppose instead, we draw this histogram with only one block covering the interval $5,000 — 15,000. How tall must this block be? c) Which is larger, the average or the median? Explain. d) The SD is closest to: i) $5,000; ii) $10,000; iii) $20,000. e) Fill in the remaining blocks with the correct percentage (as in the figure: the interval $0 — $5,000 is 15%).

Exercises

4.46 Here are the averages and SD's of five sets of numbers:

Set	Average	SD
A	60	3.0
B	5	8.0
C	5	0.2
D	20	1.5
E	40	40.0

In each data set, all the numbers are positive. The histograms of two of these sets have long right-hand tails.

a) Which two? b) Explain, using diagrams.

4.47 Draw two histograms, one with its average greater than its median and the other with its median greater than its average.

4.48 Devise two data sets: one with its average greater than its median and the other with its median greater than its average.

4.49 Here are four members of a population and their X-values.

Member	X-value
A	10
B	12
C	13
D	13

a) What is the average of X? The median? The SD?

b) Draw the histogram.

c) A fifth individual "E" enters the population. i) To increase the average by 1, what should E's X-value be? ii) To increase the range by 1, what should E's X-value be?

4.50 Estimate the average and median of each of the following histograms:

I

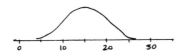

average = a) 5 b) 10 c) 15 d) 20
median = a) 5 b) 10 c) 15 d) 20

II

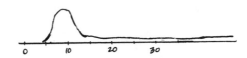

average = a) 0 b) 10 c) 15 d) 30
median = a) 0 b) 10 c) 15 d) 30

III

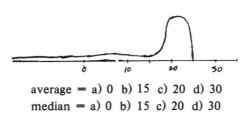

average = a) 0 b) 15 c) 20 d) 30
median = a) 0 b) 15 c) 20 d) 30

4.51 Find the average and standard deviation of the list: 4, 1, 2, 5, 2, 10.

XIII. Solutions

4.1 Average = 2

4.2 The new average is 5. Notice the answer can be obtained by direct computation, or by adding 3 to the old average: $2 + 3 = 5$.

4.3 The new average is 14. Again, this may be obtained either by direct computation or by multiplying the old average by 7: $2 \times 7 = 14$.

4.4 The average = (77 − 32 degrees Fahrenheit) × 5/9 = 25 degrees Centigrade.

4.5 The old sum = $80 \times 10 = 800$, therefore the new sum must be: $800 + 10 = 810$. So the new average is $810/10 = 81$.

4.6 The old sum was $55 \times 5 = 275$. The new sum is $275 − 63 = 212$. The new average is $212/4 = 53$ seconds.

4.7 b)

4.8 18

4.9 About 15

4.10 b)

4.11 a)

Solutions

4.12 rms = 2

4.13 rms = 3

4.14 a) Ignoring sign, these numbers have size about 5, so you should estimate the rms to be about 5. Computation shows the exact rms to be 5.2. b) Ignoring sign, the typical value in this list is about 15, so the rms will be approximately 15. Exact calculation gives 14.57. c) These numbers are, in absolute size, about 100, so 100 is a reasonable estimate of the rms. Exact calculation gives 104.5.

4.15 The easiest one to guess is c)—since, ignoring sign, all the values are 5, the rms must also be 5. So c) is 5, a) is 3.1, b) is 11.0, and d) is 15.3.

4.16 a) Average = 8; deviations = $-6, -2, 0, 2, 6$; SD = 4; b) Average = 14; deviations = $-10, -4, 4, 2, 8$; SD = 6.3.

4.17 All averages are zero: a) has SD = 1; b) and c) have SD = 2; d) through g) have SD = 3; h) SD = 8.

4.18 a) is ii); b) is iii); c) is i); and d) is iv).

4.19 a) is ii); b) is iii); and c) is i).

4.20 a)

4.21 a) False—the class might have a few extremely short people who bring the average down; b) False—consider the list: $-1, +1$. SD = 1 but average = zero; c) False—two different histograms can have the same average and SD; d) True; e) False — if the SD is zero, this merely means that all the numbers in the list are equal—they might all equal 5, for instance, in which case the average would be 5; f) True; g) False— adding 10 to each number merely shifts the histogram to the right by 10. It does not change the spread; h) False—if the right-hand tail is longer, the average lies to the right and hence is greater than the median.

4.23 c)—the average is zero: in this case the deviations from the average will be just the values in the list, and the rms and SD calculations will be identical, and d) if the rms = 0 then the SD is also zero.

4.25 d)—the sum of all the deviations must be zero. Since the first four add up to $+1$, the last one must be -1; and $6 - 1 = 5$.

4.27 One year. This is the only logical spread, since most entering freshmen were born within one year of each other.

41

4.29 SD and mode

4.31 Average = 5; mode = 3; median = 4; range = 8 - 3 = 5; SD = 2. The answer is a) and d), average and range.

4.33 b) —57%

4.35 a) 40; b) The numbers range much further above the median than below.

4.37 a) SD = 2; b) SD = 6

4.39 Average = 23.3 degrees C, SD = 3.3 degrees C

4.41 One solution is:

4.43 a) 2.5; b) i — 0.5

4.45 a) 40%; b) 4% per $1,000; c) Average is larger because the average is pulled to the right by the long right-hand tail; d) ii—an SD of $5,000 includes roughly 40% of the entries; an SD of $10,000 includes roughly 65% of the entries; an SD of $20,000 includes roughly 93% of the entries . . . roughly 68% of the entries are within one SD of the average; e) From left to right including the example: 15%, 15%, 25%, 15%, 20%, 10%.

4.47 Take any histogram with its average larger than its median:

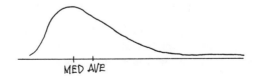

Solutions

Now reverse it and line up the median of the reversed histogram with the average of the original histogram:

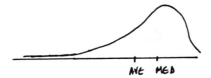

AVE   MED

4.49 a) Average = 12; median = 12.5; SD = 1.2.

b)

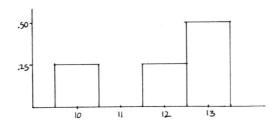

c) Seventeen increases the average by 1, 14 increases the range by 1.

4.51 Average = 4, SD = 7.3

5

The Normal Approximation for Data

I. Thinking in Standard Units

The SD is a sort of intrinsic yardstick, given to you by the data itself. It is useful for measuring the distance of a given list value from the average of the list, in terms of the spread of the list.

The idea is simple. Ask yourself how far from the average the value is (+ means above average, − means below average); then ask how many SD's this is.

Example A: The class average on a midterm is 60 and the SD is 10. You scored 75. What is your score in standard units?
Solution: Your score was 15 points above average. One SD is 10 points, so 15 points is 1.5 SD's. Your score was therefore 1.5 SD's above average—your score in standard units was +1.5.

Example B: Your midterm score in standard units is −1.0. The average score was 50 and the SD was 4. What was your score?
Solution: 1.0 SD is 4, so your score was 4 below average, or 46.

Example C: Here is summary data for two midterms:

	MT1	MT2
average	50	60
SD	10	4

Your score on the first midterm was 62 and on the second midterm your score was 64. On which midterm did you do better, relative to the class?
Solution: On the first, your score in standard units was 1.2; on the second it was 1.0. You did better, relative to the class, on the first, even though your raw score was lower.

To Convert a Number to Standard Units

Example D: Average height for men is about 68 inches and the SD is about 2.7 inches.

a) What is the height in standard units of a man 72 inches tall?

b) What is the height in standard units of a man 64 inches tall?

c) How tall is a man who is 2 SD's above average height?

d) How tall is a man who is 3 SD's below average height?

Solution: a) +1.5; b) −1.5; c) 73.4 inches; d) 60 inches

II. To Convert a Number to Standard Units

1) Find the average of the list.

2) Find the SD of the list.

3) Subtract the average from the number.

4) Divide by the SD.

Example A: A list has average = 30 and SD = 2. What is 35 in standard units?

Solution: 35 − 30 = 5, and 5/2 = 2.5. The answer is 2.5: 35 is 2.5 SD's from the average.

Example B: The LSAT exam scores average 500 with an SD of 100. You score 625. What is your score in standard units?

Solution: 625 − 500 = 125, and 125/100 = 1.25. Your score in standard units is 1.25, and this is 1.25 SD's above average.

Example C: A list has average = −50 and SD = 8. One of the numbers in the list is −40. Convert this to standard units.

Solution: −40 − (−50) = +10, and 10/8 = +1.25, so −40 in standard units is +1.25: −40 is 1.25 SD's above average.

Problems

5.1 A data set has average = 30 and SD = 10. How many SD's from average are the following values?

a) 20; b) 25; c) 30; d) 35; e) 40; f) 495; g) 50

What values are represented by the following?

h) 1.3 SD's above average; i) 0.8 SD's above average; j) 0.5 SD's above average; k) 2.5 SD's above average; l) 2.0 SD's below average

5.2 On a midterm: average = 60, SD = 8. How many SD's from the average is your score if it is: a) 68; b) 52; c) 70; d) 60; e) 50; f) 72?

What was your score if you were: g) 1.75 SD's below average; h) 2 SD's above average; i) 1.25 SD's below average?

III. Finding Areas Under the Normal Curve

The normal table gives areas for only one type of region, the region between a value and its negative. There are many possible regions, but there are only a few standard region types used in elementary statistics problems. Following is a catalogue of the most common types and the methods for calculating their areas. If you master these, you will be prepared for every situation you will meet in an elementary course. Each standard type can be assembled using "building blocks" from the normal table. You should understand this "building block" approach in each case —if you do this, you will be able to rederive the proper formula in each case as you come to it, without referring to this catalogue.

I.

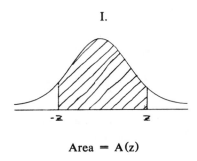

Area $= A(z)$

This area is read directly in the table—no computation is necessary.

Example A: Find the area between -0.8 and 0.8 under the normal curve.
Solution: From the normal table, $A(0.8) = 58\%$. This is the answer.

Example B: Find the area between -1.25 and 1.25 under the normal curve.
Solution: From the normal table, $A(1.25) = 79\%$.

Example C: Find the area between -0.25 and 0.25 under the normal curve.
Solution: $A(0.25) = 20\%$

Finding Areas Under the Normal Curve

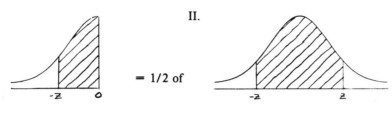

II.

= 1/2 of

Area = $(1/2) \times A(z)$

Example A: Find the area between -1.0 and 0 under the normal curve.
Solution: $(1/2) \times A(1.0) = (1/2) \times 68\% = 34\%$

Example B: Find the area between -1.4 and 0 under the normal curve.
Solution: $(1/2) \times A(1.4) = (1/2) \times 84\% = 42\%$

Example C: Find the area between 0 and 0.5 under the normal curve.
Solution: $(1/2) \times A(0.5) = (1/2) \times 38\% = 19\%$

Example D: Find the area between 0 and 1.75 under the normal curve.
Solution: $(1/2) \times A(1.75) = (1/2) \times 92\% = 46\%$

III.

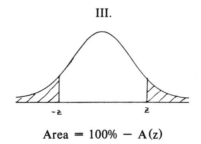

Area = $100\% - A(z)$

This is the area in the two "tails". It is what you get when you start with 100% of the curve and take away the middle part, which has area $A(z)$. Thus the area is found by taking the $A(z)$ away from 100%.

Example A: Find the total area to the left of -1.0 and to the right of 1.0 under the normal curve.
Solution: $100\% - A(1.0) = 100\% - 68\% = 32\%$

Example B: Find the area under the normal curve more than 0.6 away from zero.
Solution: $100\% - A(0.6) = 100\% - 45\% = 55\%$

Example C: Find the area under the normal curve in the two tails starting at −1.7 and +1.7.

Solution: $100\% - A(1.7) = 100\% - 91\% = 9\%$

IV.

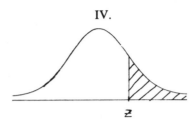

$$A = (1/2) \times (100\% - A(z))$$

This area is exactly half the area in the two tails. Pictorially,

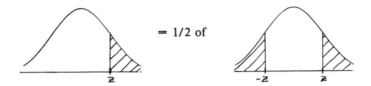

$$= 1/2 \text{ of}$$

Hence you take 1/2 of the two-tail area, which gives you:

$$(1/2) \times (100\% - A(z))$$

Example A: Find the area to the right of 0.3 under the normal curve.
Solution: $(1/2) \times (100\% - A(0.3)) = (1/2) \times (100\% - 24\%) = 38\%$

Example B: Find the area to the right of 1.25 under the normal curve.
Solution: $(1/2) \times (100\% - A(1.25)) = (1/2) \times (100\% - 79\%) = 11\%$

Example C: Find the area to the left of −1.25 under the normal curve.
Solution: $(1/2) \times (100\% - A(1.25)) = (1/2) \times (100\% - 79\%) = 11\%$

Example D: Find the area to the right of 0.75 under the normal curve.
Solution: $(1/2) \times (100\% - A(0.75)) = (1/2) \times (100\% - 55\%) = 23\%$

Example E: Find the area to the left of −0.75 under the normal curve.
Solution: $(1/2) \times (100\% - A(0.75)) = (1/2) \times (100\% - 55\%) = 23\%$

Finding Areas Under the Normal Curve

V.

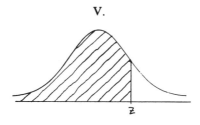

Area = 50% + (1/2) × A(z)

You can see why the area is equal to 50% + (1/2) × A(z) by the following building blocks equation:

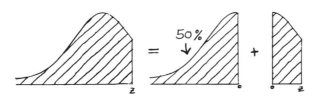

Example A: Find the area to the left of 0.95 under the normal curve.
Solution: 50% + (1/2) × A(0.95) = 50% + (1/2) × 66% = 83%

Example B: Find the area to the left of 1.55 under the normal curve.
Solution: 50% + (1/2) × A(1.55) = 50% + (1/2) × 88% = 94%

Example C: Find the area to the right of −1.0 under the normal curve.
Solution: 50% + (1/2) × A(1.0) = 50% + (1/2) × 68% = 84%

Example D: Find the area to the right of −0.3 under the normal curve.
Solution: 50% + (1/2) × A(0.3) = 50% + (1/2) × 24% = 62%

VI.

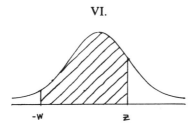

Area = (1/2) × (A(w) + A(z))

Since

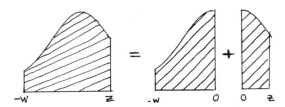

We get the area by adding the areas of these two parts. In II above we saw that the left half is $(1/2) \times A(w)$ and the right half is $(1/2) \times A(z)$.

Example A: Find the area between -1.0 and $+1.55$ under the normal curve.

Solution: $(1/2) \times A(1.0)$ + $(1/2) \times A(1.55)$ = $(1/2) \times 68\%$ + $(1/2) \times 88\% = 78\%$

Example B: Find the area between -0.5 and $+0.8$ under the normal curve.

Solution: $(1/2) \times A(0.5) + (1/2) \times A(0.8) = (1/2) \times 38\% + (1/2) \times 58\%$ $= 48\%$

Example C: Find the area between -2.0 and $+1.5$ under the normal curve.

Solution: $(1/2) \times A(2.0) + (1/2) \times A(1.5) = (1/2) \times 95\% + (1/2) \times 87\%$ $= 91\%$

Example D: Find the area between -0.3 and $+1.75$ under the normal curve.

Solution: $(1/2) \times A(0.3)$ + $(1/2) \times A(1.75)$ = $(1/2) \times 24\%$ + $(1/2) \times 92\% = 58\%$

VII.

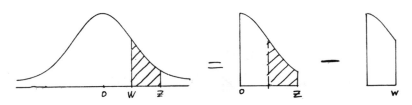

Area $= (1/2) \times (A(z) - A(w))$

The Normal Approximation for Data

Example A: Find the area between 1.0 and 1.4 under the normal curve.
Solution: $(1/2) \times A(1.4) - (1/2) \times A(1.0) = (1/2) \times 84\% - (1/2) \times 68\%$
= 8%

Example B: Find the area between 0.5 and 1.65 under the normal curve.
Solution: $(1/2) \times A(1.65) - (1/2) \times A(0.5) = (1/2) \times 90\% -$
$(1/2) \times 38\% = 26\%$

Example C: Find the area between −1.35 and −0.85 under the normal curve.
Solution: $(1/2) \times A(1.35) - (1/2) \times A(0.85) = (1/2) \times 82\% -$
$(1/2) \times 60\% = 11\%$

Example D: Find the area between −2.0 and −1.5 under the normal curve.
Solution: $(1/2) \times A(2.0) - (1/2) \times A(1.5) = (1/2) \times 95\% - (1/2) \times 87\%$
= 4%

IV. The Normal Approximation for Data

1) Draw the histogram with its scale on the horizontal axis—put the average in the center and mark the axis one SD down from the average and one SD up from the average.

2) Mark off and shade in the region of interest.

3) Convert the endpoint(s) of this region to standard units and write the standard units below the endpoint(s).

4) Use the appropriate A(z) calculation according to the above catalogue of region types.

Example A: Average LSAT score is 500 and the SD is 100. The scores follow the normal curve approximately. About what percent of the people taking the exam score above 650?
Solution:
1)

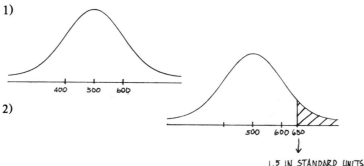

2)

1.5 IN STANDARD UNITS

51

3) $1.5 = \dfrac{650 - 500}{100}$

4) This is type IV above. The answer is:
$$(1/2) \times (100\% - A(1.5)) = (1/2) \times (100\% - 87\%) = 7\%.$$

Example B: About what percent scored between 500 and 665?
Solution: This is type II above. The answer is: $(1/2) \times A(1.65) = 45\%$.

Example C: About what percent scored between 600 and 655?
Solution: This is type VII above. The answer is:
$$(1/2) \times (A(1.55) - A(1.0)) = 10\%.$$

Example D: About what percent scored within 50 of average?
Solution: This is type I above. The answer is: $A(0.5) = 38\%$.

Example E: About what percent scored less than 640?
Solution: This is type V above. The answer is: $50\% + (1/2) \times A(1.4) = 92\%$.

Example F: About what percent scored more than 125 points away from average?
Solution: This is type III above. The answer is: $100\% - A(1.25) = 21\%$.

Example G: About what percent scored between 420 and 525?
Solution: This is type VI above. The answer is:
$$(1/2) \times (A(0.8) + A(0.25)) = 39\%.$$

Problems

5.3 Among applicants to one law school in 1973, the average LSAT score was about 600 points with an SD of about 100 points. The scores followed the normal curve. About what percentage of the applicants scored between 500 and 600 points?

5.4 The average IQ (Stanford-Binet) is 100, the SD is 16. The IQ's are normally distributed.

 a) About what percentage of the population has IQ less than 116?

 b) About what percentage of the population has IQ greater than 140?

5.5 The average score on a diagnostic quiz was 16 and the SD was 4. The scores were approximately normally distributed.

 a) About what percentage of the class scored 22 or above?

 b) About what percentage of the class scored 12 or below?

Finding Percentiles on the Normal Curve

5.6 Your final score in a class is at the 70th percentile. This means: a) 70 percent of the class did better than you; b) 30 percent of the class did better than you; c) 70 percent of the class did worse than you; d) 30 percent of the class did worse than you.

V. Percentiles

Percentiles tell you how high a number stands in a data set. If you score in the 90th percentile on a test, for instance, this means that you scored higher than 90% of the people who took the test.

Example A: You are in the 85th percentile on an exam. What percentage of the class scored higher than you?
Solution: Since 85% scored lower than you, 100% − 85% = 15% must have scored higher.

Problems

5.7 What percentage of a group of numbers lies between: a) the 30th and 70th percentiles; b) the 25th and 75th percentiles; c) the 10th and 40th percentiles; d) the 40th and 80th percentiles?

5.8 If 10 percent of a histogram is to the right of a point, the point is at the: a) 10th percentile; b) 40th percentile; c) 60th percentile; d) 90th percentile.

5.9 The 60th percentile on the histogram below is at one of the points A,B, C or D. Which one?

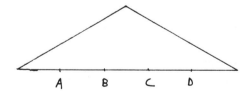

5.10 What is another name for the 50th percentile?

VI. Finding Percentiles on the Normal Curve

To find a given percentile on a normal curve you have to do three things.
1) Find the A(z)-type region that ends at the percentile.
2) Use the normal table "backwards" to get a z-value.
3) Convert the z-value to actual units.

Example A: The scores on the LSAT follow the normal curve, with an average of 500 and an SD of 100. What is the 84th percentile?

Solution: The 84th percentile on the histogram looks like this:

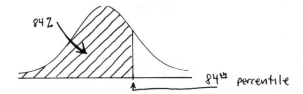

16% of the curve is to the right of the 84th percentile, so this picture converts to

1)

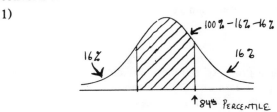

2) 100% − 16% − 16% = 68%, so looking up 68% in the A-column of the normal table, you find z = 1.0. Therefore the 84th percentile must be 1.0 in standard units.

3) 1.0 is 100 above average, so the 84th percentile is 600.

Problems

5.11 The scores on an IQ test of college graduates follow a normal curve with an average of 115 and an SD of 12.

a) A graduate in this group with an IQ of 133 has percentile rank about: i) 35%; ii) 68%; iii) 86%; iv) 93%; v) none of the above.

b) The IQ percentile rank of a college graduate in this group is 40%. His IQ score is about: i) 105; ii) 118; iii) 112; iv) between 105 and 110; v) none of the above.

5.12 Scores on the GRE among applicants to UC Berkeley are approximately normally distributed with the average at 560 and the SD at 100.

a) What percent of applicants have scores above 680?

b) 95% of the applicants have scores less than or equal to what number?

Interquartile Range

5.13 Average GPA of a college is 3.1 with an SD of 0.20. The GPA's follow the normal curve approximately. What is the: a) 84th percentile of GPA; b) 16th percentile of GPA; c) 50th percentile of GPA; d) 80th percentile of GPA?

5.14 A list of GPA's has average = 3.0 and SD = 0.4. Its histogram follows the normal curve. A student is asked to estimate the 30th percentile. She adopts a trial and error approach, first trying 2.6. Using the normal table she finds the area to the left of 2.6 is 16%. Is the 30th percentile greater than or less than 2.6?

5.15 In the normal table, what percent of the area under the histogram is between z and −z if the area to the right of z is: a) 10%; b) 20%; c) 30%; d) 40%; e) 50%?

5.16 In the normal table, approximately which value of z has: a) 20% of the histogram area between z and −z; b) 40% of the histogram area between z and −z; c) 50% of the histogram area between z and −z; d) 60% of the histogram area between z and −z; e) 80% of the histogram area between z and −z?

VII. Interquartile Range

The interquartile range is another measure of spread. This value is computed by finding the 25th and 75th percentiles of the data, then finding the distance between them. If you have the histogram, you can simply read the two scores off the histogram.

Problems

5.17 Estimate the interquartile range of the following histogram:

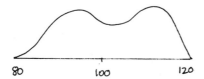

5.18 Draw a histogram with SD = 20 and interquartile range = 20.

5.19 Can you draw a histogram in which the SD is 10 times as large as the interquartile range? One hundred times?

5.20 How large can the interquartile range be relative to the SD? Can you draw a histogram with interquartile range of two SD's? Ten SD's?

VIII. Interquartile Range for the Normal Curve

If the data is normally distributed, you can calcu'ate the 25th and 75th percentiles of the data, if you know the SD. An easy number to remember in this case is 0.68—you already know that $A(1.0) = 68\%$, and if you look at the normal table you will also see that $A(0.68) = 50\%$. Therefore the 25th percentile of normal data is 0.68 SD's below the average and the 75th is 0.68 SD's above the average, as seen below.

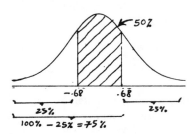

Example A: Heights of a grammar school class are normally distributed; average = 54 inches; SD = 2 inches. What is the interquartile range?
Solution: -0.68 SD's below average is $54 - 0.68 \times 2 = 54 - 1.36 = 52.64$.
0.68 SD's above average is $54 + 0.68 \times 2 = 54 + 1.36 = 55.36$.
The interquartile range $= 55.36 - 52.64 = 2.72$.

Notice that the above arithmetic is equivalent to adding 0.68 SD's to itself—thus a simple formula for the interquartile range of the normal curve is: Interquartile range $= 1.36 \times$ SD.

Problems

5.21 Exam scores are normally distributed with average = 60 and SD = 10. What is the interquartile range?

5.22 What is the MCAT interquartile range (average = 500, SD = 100)?

5.23 What is the interquartile range of Stanford-Binet IQ's (average = 100, SD = 16)?

5.24 The average on a test is 150 and the interquartile range is 68. The scores are nearly normally distributed. What is the SD?

IX. Exercises

5.25 The scores on a final exam follow the normal curve with an average of 60 and an SD of 20. The instructor wishes to give 20% A's, 30% B's, and the rest C or lower.

Exercises

a) What final score should be the cutoff between A and B?

b) What final score should be the cutoff between B and C?

5.26 Consider two normal histograms:

Histogram I has average = 72 and SD = 10.
Histogram II has average = 70 and SD = 8.

True/False

a) The 50th percentile for histogram II has the same value as the 50th percentile for histogram I.

b) since 70 = 72 − 2 and 8 = 10 − 2, the two histograms have the same spread.

c) The area to the left of 60 under histogram I is more than the area to the left of 60 under histogram II.

d) The 84th percentile for histogram I equals the 93rd percentile for histogram II.

5.27 A list of heights (in inches) is converted to standard units. The SD of the new list is: a) dependent on the average height; b) can't tell without knowing the whole list of heights; c) 1 inch; d) 1; e) normal.

5.28 For any list which is expressed in standard units the percentage of entries between 0 and 1 is about: a) 95%; b) 48%; c) 20%; d) 68%; e) can't tell without further information.

5.29 In a large lecture course, the scores on the final exam followed the normal curve closely. The average score was 60 points and 3/4 of the class scored between 50 and 70 points. The SD of the scores was: a) larger than 10 points; b) smaller than 10 points; c) impossible to say with information given.

5.30 A list of weights in pounds is converted to standard units. The SD of the new list is computed. It is: a) 0; b) 1; c) 16 lbs; d) can't tell without the data.

5.31 Here is a histogram without its horizontal scale. Its SD is 10. Below it are four scales. One of the scales belongs to the histogram. Which one?

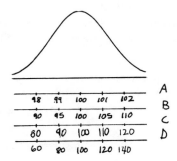

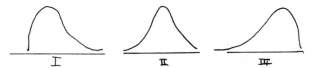

5.32 As part of a survey, one large manufacturing company asked 1,000 of its employees how far they had to commute to work each day (round trip). The data was analyzed by computer, and on the printout the average round-trip commute distance was reported as 11.3 miles, with an SD of 6.2 miles. Would a histogram for the data look like I, II, or III? Or is there a mistake somewhere?

5.33 A large lecture course is divided into two sections, each taught by a different instructor. There are 200 students in each section. Here is a summary of the results for the final exam:

Average score in first section = 56, SD = 20
Average score in second section = 63, SD = 17

About what percentage of the 400 students in the course scored better than 80 points on the final?

5.34 Which of the following variables would you expect to be not normally distributed? Explain.

a) scores on statistics final; b) income in US population; c) blood pressure in U.S. population; d) age at death in U.S. population; e) height in U.S. population

Exercises

5.35 Among applicants to one medical school in 1976, the average MCAT score was about 600 with an SD of about 100. Assume the scores followed the normal curve.

a) About what percentage of the applicants scored over 500?

b) Estimate the 90th percentile for the scores.

5.36 Two of the following data sets are far from normally distributed. On the basis of the summary statistics given, decide which two and draw a histogram for each of the two.

	Average	SD	median
A:	20	5	22
B:	−10	10	−8
C:	100	1	98
D:	20	0.3	22
E:	0	1	−0.1

5.37 Family income is not normally distributed: the histogram has a long right-handed tail. Here is a summary of family income data from 10,000 families in a certain town:

average income = $28,000/year, SD income = $10,000/year

The normal approximation says that the percentage of families with less than average income should be about 50%. Knowing that this data has a long right-hand tail, would you expect the actual percentage of below-average families to be greater than or less than 50%?

5.38 Height in a population is approximately normally distributed, with an average of about 68 inches and an SD of about 3 inches. Consider those people in the population who are over 71 inches tall.

a) Is height in this group normally distributed?

b) What can you say about the average height of the group? The SD?

5.39 The ETS Verbal Aptitude Test is designed so that the scores of high school seniors taking the test will have a normal distribution with average 500 and SD 100.

a) What percentage of students' Verbal Aptitude scores are between 450 and 550?

b) A college wants their students to be in the top 12.5 percent for verbal aptitude. What minimum qualifying score should they set?

5.40 Scores on the LSAT follow the normal curve with an average of 500 and an SD of 100. Consider the group of people who take the LSAT and get scores of 600 or more. What percentage of this group actually had scores of 650 or more?

5.41 In 1974, men averaged about 500 on the mathematical SAT, while women averaged about 460. Both histograms followed the normal curve, with an SD of about 100.

a) Estimate the percentage of men getting over 650 on this test in 1974.

b) Estimate the percentage of women getting over 650 on this test in 1974.

5.42 Below is a table of blood pressures of subjects in a study. The table gives the density over each interval. Average blood pressure for the group is 120 mm, SD is 10 mm.

Interval (mm)	%/mm
90–100	0.4
100–105	1.0
105–110	2.0
110–115	3.0
115–120	3.4
120–125	2.6
125–130	2.4
130–135	2.0
135–140	1.0
140–150	0.6
150–160	0.2

a) What is the normal approximation to the percentage of subjects with blood pressure between 110 and 130 mm?

b) What is the actual percentage of subjects with blood pressure between 110 mm and 130 mm?

c) How do you explain the difference?

5.43 Can you sketch an approximately normal curve in which: a) the median and average are both larger than the SD; b) the average is much larger than the median; c) the median is smaller than the SD; d) the average is smaller than the SD and the data consists entirely of non-negative numbers?

Exercises

5.44 In a large study of the growth of children, the following results were obtained:

> For boys aged 12: average height = 58.5 inches, SD height = 2.5 inches
>
> For boys aged 14: average height = 63.75 inches, SD height = 2.5 inches

a) Sketch the histogram for the twelve-year olds, labeling the average. Then label the histogram with the value of the average height for the fourteen-year-olds. Finally, shade the area to the right of this value.

b) About what percent of the 12-year-olds in the study were taller than the average height for the 14-year-olds?

5.45 The median score on a test was 24 points (out of 50 points). The histogram of the scores follows the normal curve closely. Someone you know tells you he scored 34 points and ranked ahead of 84% of the class. You scored 39 points. What is your rank?

5.46 Thoroughbred race horses of a certain class at a New York racetrack run six furlongs in various times. The average time is 72 seconds with an SD of one second. Consider those horses at this track that run six furlongs in 71 seconds or less. What percentage of this group would you expect to run six furlongs in 70 seconds or better, if the running times were normally distributed?

5.47 Find the area under the normal curve: a) outside (-1.5) and (1.5); b) to the right of (-0.8); c) between (-0.3) and (0.9); d) between (-2.5) and (2.0); e) between (-0.386) and (1.309).

5.48 Here is a histogram:

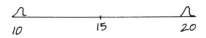

a) What is the average?
b) What is the SD?
c) What is the interquartile range?

5.49 The figure below is a histogram for the scores on the final in a class. Find the 75th percentile of these scores.

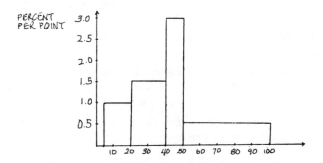

5.50 The figure below is a histogram for the scores on the final in a class. Find the 70th percentile of these scores.

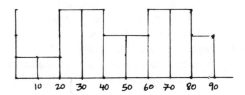

5.51 Of those admitted to law school, 80% have scored 600 or more on the LSAT. Of those rejected by law school, 75% have scored less than 600. Of those scoring 600 or more on the LSAT, what percentage get admitted to law school?

a) 80%; b) 75%; c) 20%; d) 25%; e) cannot be determined from the data given

5.52 A person's score is at the 90th percentile in a population.

a) If the population is normal with average 500 and SD 100, what is his score?

b) In the same population, a person scoring 300 is at what percentile?

5.53 A data set is normally distributed with average = 300 and SD = 10. What is the interquartile range?

Solutions

5.54 The figure below is a histogram for the scores on the final in a certain class. Find the 50th percentile of these scores.

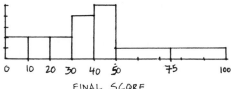

FINAL SCORE

5.55 In a town, it was found that the average family size was 2.8, with an SD of 3.0.

a) Is the percentage of families with family size greater than 2.8 more than 50% or less than 50%? Explain.

b) Can you use the normal approximation for this histogram? Explain.

5.56 A data set is normally distributed with interquartile range = 30. Is it possible to determine the average or SD from this information alone? Why or why not?

5.57 You are looking at a computer printout of 100 numbers which have been converted to standard units. Looking at the first ten entries, you see:

$$-6.2, \ 3.5, \ 1.2, \ -0.13, \ 4.2, \ -5.1, \ -7.2, \ -11.3, \ 1.8, \ 6.3$$

Is anything wrong?

X. Solutions

5.1 a) -1; b) -0.5; c) 0; d) 0.5; e) 1.0; f) 1.5; g) 2.0; h) 43; i) 38; j) 35; k) 55; l) 10

5.2 a) 1.0; b) -1.0; c) 1.25; d) 0; e) -1.25; f) 1.5; g) 46; h) 76; i) 50

5.3 This is type II; $(1/2) \times A(1) = 34\%$.

5.4 a) This is type V; $50\% + (1/2) \times A(1) = 84\%$; b) This is type IV; $(1/2) \times (100\% - A(2.5)) = 0.5\%$

5.5 a) This is type IV; $(1/2) \times (100\% - A(1.5)) = 7\%$; b) This is type IV; $(1/2) \times (100\% - A(1)) = 16\%$.

5.6 b and c

5.7 a) 40% (= 70% − 30%); b) 50% (= 75% − 25%); c) 30% (= 40% − 10%); d) 40% (= 80% − 40%)

5.8 d—since 10% is to the right, the rest—or 90%—must be to the left and the point is therefore by definition the 90th percentile.

5.9 C

5.10 The median

5.11 a) iv—93%; b) iii—112

5.12 a) About 12%; b) 725 (z = 1.65)

5.13 a) 84th percentile has z = 1.0, so it is 3.1 + 0.20 = 3.30; b) 16th percentile has z = −1.0, so it is 3.1 − 0.20 = 2.90; c) 50th percentile has z = 0, so it is 3.1; d) 80th percentile has z = 0.85, so it is 3.1 + 0.85 × 0.20 = 3.27.

5.14 Greater than 2.6, since 30% of the histogram must be to the left of the 30th percentile and only 16% is to the left of 2.6.

5.15 a) 80%—since also 10% of the area is to the left of −z, and the area in the middle is found by subtracting the 10% in each tail: 100% − 10% − 10% = 80%; b) 60% = 100% − 20% − 20%; c) 40% = 100% − 30% − 30%; d) 20% = 100% − 40% − 40%; e) 0%, since z = 0 and −z therefore also equals zero.

5.16 a) 0.25, since A(0.25) = 20%; b) 0.525, since A(0.50) = 38%, and A(0.55) = 42%, and 40% is about halfway between 38% and 42%; c) 0.68, by interpolation between A(0.65) = 48% and A(0.70) = 52%; d) 0.85, since A(0.85) = 60%; e) 1.30, since A(1.30) = 81%.

5.17 25th percentile appears to be about 90 and the 75th is about 110 giving an interquartile range of about 110 − 90 = 20.

5.18

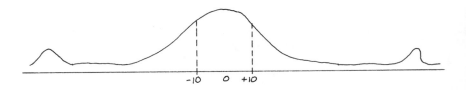

Solutions

5.19 Yes. Just move the two lumps in the solution to the last problem further away from the average. This will increase the SD, but will not change the interquartile range. These two problems illustrate the fact that the interquartile range, like the median, is not sensitive to outliers.

5.20 If you try to draw histograms with interquartile ranges much larger than their SD's, you will quickly see that there is a limit: the interquartile range cannot be very much larger than one SD.

5.21 $1.36 \times 10 = 13.6$

5.22 $1.36 \times 100 = 136$

5.23 $1.36 \times 16 = 21.76$

5.24 Since $68 = 1.36$ SD's, one SD must be 50.

5.25 a) 77; b) 60

5.27 d

5.29 Smaller than 10 points. Since the average was 60, about 2/3 of the class had scores within one SD of 60. But even more than that—3/4 had scores between 50 and 70—that is, within 10 of 60. So 10 must be more than one SD.

5.31 The histogram is approximately normal with average $= 100$; therefore the area between 90 and 110 must be about 68%. Only scale C achieves this.

5.33 14%—in the first section, 80 is 1.20 SD's above average, so about 12% of the 200—or 24 students—scored better than 80. In the second section, 80 is 1 SD above average, so about 16% of this group—or 32 students—scored better than 80. So altogether $32 + 24 = 56$ students scored above 80 and 56 is 14% of 400.

5.35 a) 84%; b) 730

5.37 Greater than 50%, to balance the long right-hand tail.

5.39 a) $A(0.5) = 38\%$; b) 1.15 SD's above average, which is 615.

5.41 a) $(1/2) \times (100\% - A(1.5)) = 6.5\%$; b) $(1/2) \times (100\% - A(1.9)) = 3\%$.

5.43 a)

b) No—in normal data, the average and the median must be about the same.

c)

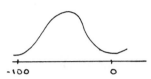

d) No—in normal data, about 16% of the histogram is more than one SD below the average. If an SD is larger than the average, this places at least 16% of the data below zero.

5.45 Since the scores follow the normal curve, the average is the same as the median, 24. Since 34 is the 84th percentile and the 84th percentile corresponds to 1.0 SD's above average, $34 - 24 = 10 = 1$ SD. So you are 1.5 SD's above average, which is the 93.5th percentile.

5.47 a) $100\% - A(1.5) = 13\%$; b) $50\% + (1/2) \times A(0.8) = 79\%$; c) $(1/2) \times A(0.3) + (1/2) \times A(0.9) = 43.5\%$; d) $(1/2) \times A(2.5) + (1/2) \times A(2.0) = 97\%$; e) $(1/2) \times A(0.386) + (1/2) \times A(1.309) = 56\%$

5.49 50

5.51 e

5.53 $1.36 \times 10 = 13.6$

5.55 a) The histogram evidently has a long right-hand tail—since family size must be at least 1 and the SD is larger than the average. So the median lies to the left of the average, and hence less than 50% lies to the right of the average. b) No—its long right hand tail will make the normal approximation fail.

5.57 Yes—the list should contain only about 5% values greater than 2.0 but this list has 6 out of 10 greater than 2.0.

6

Measurement Error

Example A: On page 91 (FPP) notice that all measurements of NB 10 are smaller than 10 grams.

a) Can you conclude that NB 10's actual weight is less than 10 grams?

b) What would be a reasonable way to estimate the actual weight of NB 10?

Solution: a) No but if it isn't the measuring device is very biased—it is not likely that the chance error would be this negative every time.

b) Compute the average of all the measurements—in computing the average the positive and negative errors would tend to cancel one another out.

7

Graphing

I. Reading Points Off a Graph

You should be able to describe the characteristics represented by a point on a graph.

Example A: Here is a graph of five people's GPA/LSAT scores. The x-coordinate is GPA and the y-coordinate is LSAT.

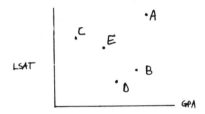

a) Who had the highest GPA? b) Who had the lowest GPA? c) Who had the highest LSAT? d) Who had the second highest LSAT? e) Who had the lowest LSAT?

Solution: a) A; b) C; c) A; d) C; e) D

Example B: Here is a graph of five people's IQ/GPA measurements. The x-coordinate is IQ and the y-coordinate is GPA.

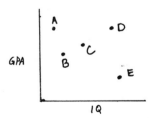

Slope and Intercept

Of the 5, 3 can be described as follows—who fits the descriptions?

(i) bright but lazy; (ii) very hard worker; (iii) bright and a good student too

Solution: A has the lowest IQ but the highest GPA—A is evidently a very hard worker (ii); E has the highest IQ but the lowest GPA—E is bright but lazy (i); D has a high IQ and also has a high GPA—D is bright and a good student too (iii).

Problem

7.1 Here is a scatter diagram of five people's height/weight measurements. They can be described as follows. Who fits which description?

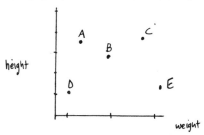

(i) short and thin; (ii) short and heavy; (iii) tall and thin; (iv) tall and heavy; (v) about average height and weight

II. Slope and Intercept

Likewise, you should have a good geometrical idea of what slope means. You should, in particular, be able to estimate slope and intercept from a brief inspection of the graph.

Example A: Below is a graph with five lines.

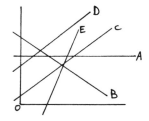

a) Which line has the greatest slope? b) Which line(s) have negative slope? c) Which line(s) have negative intercept? d) Which line has the greatest intercept? e) Which line has the smallest intercept? f) Which lines have the same slope? g) Which lines have the same intercept?

Solution: a) E; b) B; c) only E; d) B; e) E; f) C and D; g) none—the intercepts are all different

Problems

7.2 Match the lines on the graph with their correct slope/intercept pairs.

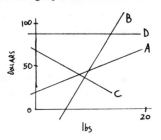

slope ($/lb.)	intercept ($)
1.5	20
-1.5	60
0.0	90
6.0	-25

7.3

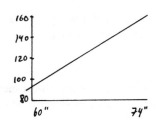

a) In the diagram, the slope of the line is: i) 1; ii) 2; iii) 3; iv) 4.

b) The intercept is one of the following—which one? i) 80; ii) 90; iii) 100; iv) 110; v) 120.

III. Solutions

7.1 A—iii; B—v; C—iv; D—i; E—ii

7.2 The only negative intercept is B, so B must be (6,-25). The only slope of 0 is D, so D must be (0,90). The only negative slope is C, so C must be (-1.5,60). A is then the only one left, and it must be (1.5,20).

7.3 a) iv; b) ii

8

Correlation

I. Choosing the Dependent Variable

In any study of two related variables, one of the variables often plays the role of dependent variable more naturally than the other. The vertical axis on the scatter diagram is usually assigned to this variable. The dependent variable should have one of the following characteristics.

1) Its measurement comes after the other's in time.

2) It is harder to measure than the other.

3) It depends in some physical way upon the other.

4) It is the output of a system in which the other variable is the input.

5) It is the variable whose value one would like to predict, using the values of the other variable.

Example A: In a study of LSAT score and law school GPA, which is the natural dependent variable?

Solution: LSAT score is measured first, law school GPA comes afterward—so law school GPA is the natural dependent variable.

Example B: In a study of college GPA and LSAT scores, which is the natural dependent variable?

Solution: In this case, LSAT scores come after college GPA, so LSAT score is the natural dependent variable.

Example C: The highway patrol uses a field instrument to measure approximate alcohol content in the blood, in order to decide whether a driver is legally drunk. This use is founded upon correlation studies relating alcohol level measurement on this instrument to true alcohol measurement as determined by a hospital laboratory. In such a study, which is the natural dependent variable?

Solution: It is more difficult to measure true alcohol content in the blood—so this is the dependent variable.

Problems

8.1 A correlation study relates the compression (in inches) of a shock absorber to the weight placed upon it. Which is the natural dependent variable?

8.2 A correlation study relates the productivity of factory workers to the amount of bonus money awarded to them. Which is the natural dependent variable?

8.3 An economist attempts to predict the price of sugar by studying the relationship between supply-demand imbalance and sugar prices one year hence. Which is the natural dependent variable?

II. Locating the Average and SD on a Scatter Diagram

The way to locate the average (or SD) of one of the variables on a scatter diagram is to imagine all the points collapsed down to form a histogram on that axis. Then use your knowledge of histograms to locate the average (or SD) of the histogram—that is the average (or SD) of that variable.

Example A: What is the average of x in the following scatter diagram?

Solution: Collapsing the points onto the x-axis gives the following picture:

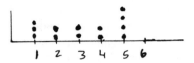

It appears the average is about 4. Calculation shows it to be exactly:

$$\frac{3 \times 1 + 2 \times 2 + 2 \times 3 + 2 \times 4 + 4 \times 5}{13} = \frac{41}{13} = 3.154$$

Estimating the Correlation Coefficient from a Scatter Diagram

Example B: Estimate the average of y in the following scatter diagram:

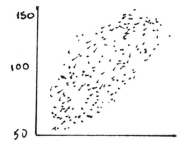

Solution: It is impractical to collapse all the points onto the y-axis. However you can see what you would get as an average if you did—you would get a roughly normal curve centered at 100.

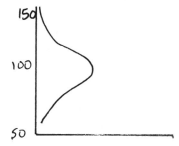

Thus the average of y is about 100.

Problem

8.4 Estimate the SD of y in the above scatter diagram.

III. Estimating the Correlation Coefficient from a Scatter Diagram

You should get acquainted with the scatter associated with various correlation coefficients. The only way to do this is by looking at a number of scatter diagrams. Those on pp. 119 and 121 in FPP are a start. Here are some others. To guess the correlation of a new scatter diagram, ask yourself which of the scatter diagrams on pp. 119 and 121 most resembles it in the scatter of its points.

Example A: Guess the correlation coefficient of the following:

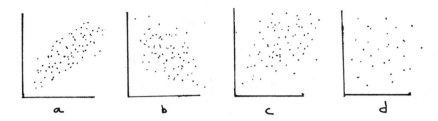

Solution: a) this is about 0.8; b) this is about -0.5; c) this is about 0.4; d) this is near zero.

Problem

8.5 Which of the following two data sets has a larger correlation coefficient?

Data Set One

x:	1	1	-1	-1	5
y:	1	-1	1	-1	5

Data Set Two

x:	1	1	-1	-1	10
y:	1	-1	1	-1	10

IV. Making Judgements About the Correlation Coefficient

You should also be able to make intuitive judgements about the correlation between related measurements of objects well known to you.

Example A: True/False

You would expect the age of a ping-pong ball and the height of its bounce to be negatively correlated.

Solution: True—the older the ping-pong ball, the less bounce it has—so older balls should not bounce as high as new ones.

Problem

8.6 Would you expect the correlation between weight of a car and miles per gallon to be positively or negatively correlated?

Exercises

V. Computing the Correlation Coefficient

Use the following program to compute the correlation coefficient:

1) Convert the x-values to standard units.

2) Convert the y-values to standard units.

3) Multiply each x-value (in standard units) by each corresponding y-value.

4) The correlation coefficient (r) is the average of the products.

8.7 Find the correlation coefficient for each of the two data sets below:

Data Set I

x:	1	1	2	2	4	2
y:	3	3	5	2	2	3

Data Set II

x:	1	2	4	1	2	2
y:	4	5	7	4	5	5

8.8 Find the correlation coefficient of the following data:

x:	12	14	25	32	40
y:	-3	-1	10	17	25

8.9 Find the correlation coefficient of the following data:

x:	4	4	3	3	1	3
y:	4	2	1	5	5	7

VI. Exercises

8.10 True/False

a) If you have the scatter diagram for a data set, you can derive the histogram of each variable from it.

b) If you have two histograms, you can construct the scatter diagram for the two variables.

c) A study of a college group shows that the correlation between GPA freshman year and GPA sophomore year is 0.5. You can conclude that most students have a higher GPA their sophomore year.

d) Scatter diagram A has $r = 0.3$, and scatter diagram B has $r = 0.4$. Scatter diagram C is created by combining data from A and B. The correlation coefficient of diagram $C = 0.7$.

e) If the correlation coefficient is -0.8, you expect negative values of the dependent variable to be associated with negative values of the independent variable.

8.11 For a group of bicyclists commuting to campus on city streets on a given day, the correlation between *time spent waiting at traffic signals* and *total cycling time* was calculated at 0.50.

True/False

a) The average bike rider spent 50% of his cycling time waiting at traffic signals.

b) If a cyclist's total riding time increases by 10 minutes he will spend an additional 5 minutes waiting at traffic signals, on the average.

c) The more time a bike rider (in the group) spends waiting at traffic signals, the longer is his total riding time, on the average.

d) The more time a bike rider (in the group) spends commuting to campus, the more time he spends waiting at traffic signals, on the average.

8.12 For which of the following pairs of variables does it make sense to use the correlation coefficient as a measure of strength of relationship?

a) IQ and salary; b) Sex and salary; c) Sex and weight; d) Weight and age

8.13 An investigator wants to study the relationship between physical and intellectual growth of children. He has data which give the size of vocabulary and the height of all children in a large elementary school. Within each of the grades, 1 through 6, the correlation between height and size of vocabulary is about zero. For all students in the school, the correlation is likely to be: a) about zero; b) positive; c) negative; d) no way to tell.

Exercises

8.14 Here are six scatter diagrams:

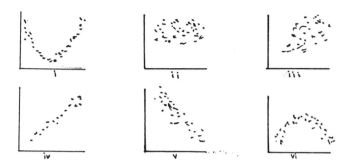

Independent Variable	Dependent Variable
a) Age of a tennis ball	Height of its bounce
b) Age of a child	Height of the child
c) Height of an adult	Weight of the adult
d) Age of an adult	Height of the adult

Match each diagram with its data set.

8.15 A study is conducted relating total current construction costs of single-family dwellings in all areas of Oakland to the number of square feet of living space. The correlation coefficient is computed. It is closest to: a) -1.0; b) -0.3; c) 0; d) 0.3; e) 1.0.

8.16 Four data sets are collected and the correlation coefficient in each case is computed. The data sets are: a) GPA freshman year vs. GPA sophomore year; b) Weight of automobile vs. miles/gallon on highways; c) GPA freshman year vs. GPA senior year; d) Length of a 2 × 4 board vs. its weight.

The correlations (scrambled) are: -0.6, 0.4, 0.7, 0.9. Match the correlations with the data sets.

8.17 The correlation between the ages of husbands and wives in the U.S. in 1976 was one of the following. Which?

a) -1.0; b) -0.92; c) 0.92; d) 1.0; e) cannot tell without further data

8.18 Match each of the following data sets with the correct correlation coefficient.

a) Data: For each bag of pennies in a bank vault; x = number of pennies in the sack, y = weight of the sack.

b) Data: On a trip from Berkeley to Los Angeles, make several measurements of the distance to both cities. Let x = distance from Berkeley and y = distance to Los Angeles.

c) Data: Observe 100 cars from the street. Let x = gas consumption in miles/gallon and y = weight of car.

d) Data: For all juniors at Univ. of California at Berkeley; x = GPA sophomore year, y = GPA junior year.

Correlation coefficients: i) -1.0; ii) -0.6; iii) 0; iv) +0.6; v) +1.0

8.19 Suppose there are two scatter diagrams with the same point of averages and the same SD's for both the independent and dependent variable. The first scatter diagram consists of 50 points and has a correlation coefficient of -.06. The second scatter diagram consists of 100 points and has a correlation coefficient of +0.6. The correlation of the combined data is: a) less than zero; b) equal to zero; c) greater than zero; d) can't say.

8.20 Which of the following pairs of variables can be studied using correlation? a) height vs. income; b) height of wife vs. height of husband; c) hair color of wife vs. hair color of husband; d) height of industrial chimney vs. amount of pollution in immediate neighborhood; e) religion vs. family size; f) amount of body fat vs. average number of hours of jogging per day; g) age vs. annual income; h) race vs. annual income

8.21 For each of the scatter diagrams below, state whether the correlation coefficient is: a) 0; b) less than 0; c) between 1 and 0; d) 1.

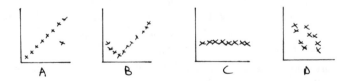

8.22 A large group of children, aged 6 to 11, had their heights and weights measured. At each age, the correlation coefficient was about 0.5. For all the children together, would the correlation be: a) somewhat more than 0.5; b) about 0.5; c) somewhat less than 0.5?

8.23 If the correlation between x and y is negative, then as x gets larger, y gets: a) negative, on the average; b) smaller, on the average; c) negative and smaller, on the average.

Exercises

8.24 A study of the IQ's of husbands and wives yielded the following:

For husbands, average IQ = 100 points; SD = 15 points.
For wives, average IQ = 100 points; SD = 15 points.
correlation = +0.6.

One of the following is a scatter diagram for the data. Which one?

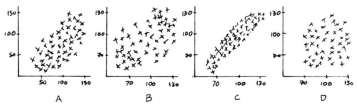

A   B   C   D

8.25 For each scatter diagram below:

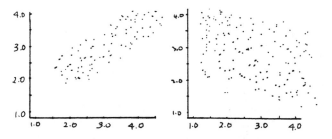

a) Guess whether the average of x is closest to: 1.0, 1.5, 2.0, 2.5, 3.0, 3.5 or 4.0; b) Do the same for the average of y; c) Guess whether the SD of x is closest to: 0.25, 0.5, 1.0, or 1.5; d) Do the same for the SD of y; e) Is the correlation positive, negative, or zero? f) For which diagram is the correlation closest to zero, ignoring sign?

8.26 Four people have their weight and height measured in pounds and inches. The results are:

Inches	Pounds
62	103
70	121
69	116
72	124

a) Find the averages and the SD's; b) Draw a scatter diagram; c) without computing the correlation between height and weight, which of the following does your diagram suggest it to be: i) negative, ii) near zero; iii) about 0.3; iv) 1.0; v) near 1, but not quite?

8.27 While the Three Mile Island reactor was experiencing its problem, a series of readings of the reactor's temperature and pressure were recorded, as shown in the scatter diagram below.

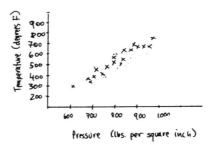

a) Sketch the SD line.

b) The SD of the temperature values is about _____.

c) The correlation coefficient is closest to: i) 0.25; ii) 0.5; iii) 0.9.

d) Suppose you are told that temperature and pressure are related by the equation: temperature = 1/pressure. Does this equation agree with what the scatter diagram tells you?

8.28 At one law school, the correlation between LSAT score and GPA for first-year students was 0.4. Average LSAT of those admitted was 650 with an SD of 50; average GPA was 3.0 with an SD of 0.4. Both scores approximately follow the normal curve. Draw a scatter diagram to help answer the following questions:

a) Approximately what percentage of all students admitted was above average on both the LSAT score and law school GPA? b) At another school, the averages and SD's are the same, but the correlation is 0.8. Is the percentage of students above average in both areas higher or lower than at the first school? c) What is the highest that this percentage could be at a third law school? What would be the corresponding correlation? d) What is the lowest this percentage could be? What would be the corresponding correlation?

8.29 A teaching assistant gives a quiz in his section. There are ten questions on the quiz and no partial credit is given. After grading the papers the TA writes down for each student the number of questions the student got right and the number he got wrong. The average number of correct answers is 6.4 with an SD of 2.0.

a) What is the average and SD of the number of wrong answers?
b) The correlation coefficient between the numbers of right and wrong answers is: i) 0; ii) -0.5; iii) +0.5; iv) -1; v) +1; vi) can't tell without the data.

8.30 A factory has twenty machines producing a product. Usually there are a number of machines that are out of order. What is the correlation coefficient between the number of machines that are out of order and the number which are working? Explain.

8.31 Five people have the following ages: 10, 11, 10, 14 and 15.

a) find the average and the SD. b) These people take a test, with possible scores between 0 and 100. It is found that the correlation between age and test score is between 0.4 and 0.6. Draw a possible scatter diagram showing age and test score for each individual.

8.32 If women always married men who were five years older, what would the correlation between their ages be? Why?

8.33 As part of a study on healthy families, two raters, A and B, looked at videotapes of 23 families, and rated the "family strength" of each on a scale of 1 to 10. Each family was rated separately, once by A and once by B. Because of unforeseen difficulties, the rating session took eight hours to complete, and the raters seemed to tire in the last half of the session.

The investigators reported a correlation of .75 between A and B's ratings for the whole session. Since they were concerned about the fatigue effect, they divided A and B's work into two groups, those taken during the first half of the evaluation and those taken during the second half.

The results: A and B correlated 0.70 for the first four hours, but only 0.26 during the last four hours. The resulting correlation for the whole session was larger than both the correlation for the first and for the second. The investigators did not comment on this. Did they make a mistake somewhere, or is there a possible explanation?

8.34 For each of the following pairs of measurements, decide whether the correlation is positive or negative and make a reasonable estimate of the correlation coefficient:

a) percent of body fat vs. average number of hours jogging per day; b) size of automobile engine in cc's vs. miles per gallon; c) number of students at college vs. number of textbooks sold in the campus bookstore; d) height vs. income of executives; e) educational level of parents vs. educational level of children; f) number of people using a ski slope each day vs. number of injuries on that slope that day;

g) number of children in a family vs. number of family automobiles;
h) hours of training per day vs. time in 100-yard freestyle; i) square footage vs. market price of house

8.35 Rank in order—from highest to lowest—the correlations between:
a) purchase price of a home and yearly income of purchasing family.
b) purchase price of a home and shoe size of the purchaser.
c) asking price and selling price of homes in the San Francisco Bay Area.
d) purchase price of a home and age of the buyer.
e) purchase price of a home and age of the home.

8.36 a) Plot the data set given below on a scatter diagram.

x:	0	1	3	4	4	6
y:	2	1	1	2	4	2

b) Calculate the correlation coefficient.

8.37 After receiving a dose of a certain drug, each of three individuals has his temperature and blood pressure measured. The results are:

Temperature	Blood Pressure
99	100
97	80
104	150

a) Find the averages; b) Find the SD's; c) Use these figures to find the correlation between blood pressure and temperature.

8.38 For each of five individuals, measurements on variables U, V, and W are given below:

U	V	W
61	14	144
60	20	204
57	22	224
63	18	184
59	26	264

Compute: a) The average of each variable; b) The SD for each variable; c) The correlations between U and V, V and W, U and W.

Exercises

8.39 a) Find the correlation coefficient for the data below:

x: 0 0 1 1 1 3
y: 1 2 2 2 4 7

b) Would the SD of the y-values change if the "4" and "7" were interchanged? Would the correlation coefficient change?

8.40 Below is a scatter diagram for data from a height-weight study. The vertical strip includes all subjects who were 72 inches tall to the nearest inch.

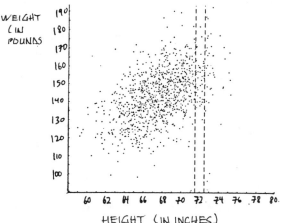

a) How much did the heaviest person in the group weigh? b) How much did the lightest person in the group weigh? c) Looking at the picture, estimate the average weight of the people in this group. d) About what percentage of the people in this group weighed more than 132 pounds?

8.41 Find the correlation coefficient for the data set below:

x: 9 9 8 8 6 8
y: 8 4 2 10 10 14

8.42 A key punch operator punched the heights and weights of 20 people on cards. Each card had one individual's height and weight, height first. These cards were then used to calculate the correlation between height and weight.

a) Suppose the cards were dropped and when placed in a computer they were in a different order. Would this change the correlation coefficient? b) Suppose the operator interchanged the height and weight on one card. Would this effect r? c) Suppose the operator made a mistake and typed a "2" instead of a "1" on the first person's height. Would this mistake cause the correlation coefficient to differ? d) Suppose the operator lost one card. What would happen to the correlation coefficient: would it change or not?

VII. Solutions

8.1 The amount of compression depends upon the weight—so it is the dependent variable.

8.2 The bonus money is thought of as input to the worker system, where the output is productivity. So productivity is the dependent variable.

8.3 The price of sugar is being predicted; it also comes later than the supply-demand imbalance—so the price of sugar is the dependent variable.

8.4 Looking again at the histogram we derived, the histogram's SD appears to be about 20 or so. Twenty is a reasonable estimate of the SD of y.

8.5 Data Set Two

8.6 Negatively correlated—the heavier the car, the more gas it burns and the lower the miles per gallon.

8.7

Data Set I		x in standard units	y in standard units	Products
x	y			
1	3	-1	0	0
1	3	-1	0	0
2	5	0	2	0
2	2	0	-1	0
4	2	2	-1	-2
2	3	0	0	0

The correlation coefficient for Data Set I is: $\frac{-2}{6} = -0.3$.

Solutions

Data Set II		x in standard units	y in standard units	Products
x	y			
1	4	-1	-1	1
2	5	0	0	0
4	7	2	2	4
1	4	-1	-1	1
2	5	0	0	0
2	5	0	0	0

The correlation coefficient for Data Set II is: $\frac{6}{6} = +1.0$. Notice that y = x + 3, i.e. this is a perfect linear relationship.

8.8

Data	
x	y
12	-3
14	-1
25	10
32	17
40	25

Here $r = +1.0$. This also could be seen by plotting the points and seeing that they lie in a straight line, or by noticing that y = x + 15.

8.9

Data		x in standard units	y in standard units	Products
x	y			
4	4	1	0	0
4	2	1	-1	-1
3	1	0	-1.5	0
3	5	0	0.5	0
1	5	-2	0.5	-1
3	7	0	1.5	0

Here $r = \frac{-2}{6} = -0.33$.

8.11 a) False; b) False; c) True; d) True

8.13 Positive. As children go from grade 1 to 6, height and vocabulary increase.

8.15 e)

8.17 c)—the correlation must be positive and high, but it can't be +1.

8.19 c)

8.21 A is between 1 and 0; B is between 1 and 0; C is 0; D is less than 0.

8.23 b)

8.25 a) 3, 2.5; b) 3.0, 2.5; c) 0.5, 0.5; d) 0.5, 0.5; e) positive, negative; f) II

8.27 a)

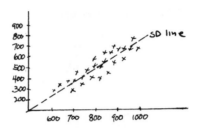

b) 130; c) iii-0.9; d) no

8.29 a) Since the average number of correct answers is 6.4, the average number of incorrect answers must be 10 - 6.4 = 3.6. The SD of the number of correct answers is the same as the SD of the number of incorrect answers: 2.0; b) iv. There is a perfect linear relationship.

8.31 a) Average = 12, SD = 2.1
b)

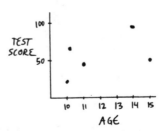

8.33 No mistake—A & B may have given very low (or very high) ratings during the last four hours because their fatigue made them unwilling (or unable) to make fine discriminations.

Solutions

8.35 a) must be farily high because of bank financial requirements, but there is great variation in purchase prices of homes among high income families, which leads to a correlation statistically less than 1.0. a) is lower than c); b)any significantly positive or negative correlation between purchase price and shoe size is most unlikely; c) is probably close to 1.0, since selling price is almost always very close to asking price. To see this, draw a scatter diagram with some typical points; d) is definitely positive, but probably not very strong; e) is negative, athough possibly not very negative. Thus the correlations are, in order, as follows: c), a), e), d), b).

8.37 a) Averages are 100 and 110; b) SD's are 2.9 and 29.4; c) r $= +1.0$

8.39 a) $+0.92$; b) SD wouldn't change but r would.

8.41 a) r $= -0.3$

9

More About Correlation

I. Exercises

9.1 True/False

For each of five Statistics classes, the average percent correct on the midterm and the average percent correct on the final for the class were calculated. The correlation for the five pairs of averages was $+0.97$. Does this mean that the relation between a student's midterm and final scores for students in the five classes is almost exactly a line?

9.2 True/False

After being given a dose of a certain drug, the average temperatures and blood pressures of five groups of 20 subjects were recorded. The correlation for the five pairs of averages was $+0.92$. The relation between a subject's temperature and blood pressure in the group is almost exactly a line.

II. Solutions

9.1 No; this is an ecological correlation—if you look at the raw (unaveraged) data, the points will be much more dispersed and the correlation will be lower.

10

Regression

I. Computing the Regression Estimate

In this type of problem you are always given two variables and five summary statistics: two averages, two SD's and a correlation. Here is an example:

Example A: In a population the average height is 68 inches and the SD of heights is 3 inches while the average weight is 120 lbs. and the SD of weight is 151 lbs. The correlation between weight and height is 0.6.

It is natural to ask questions about particular subpopulations of the group under study. The subpopulations that are easily described using this data are those having a given value for one of the measurements. For instance:

Example B: 1) Those whose height is 68 inches.

2) Those whose height is 62 inches.

3) Those whose weight is 125 lbs.

4) Those whose weight is 118 lbs.

The standard question you learn to answer in this chapter is: "How can you describe these subgroups' *other* measurement"?

Example C:

1) About how much do those weigh who are 68 inches tall?

2) About how much do those weigh who are 62 inches tall?

3) How tall on average are those people who weigh 125 lbs?

4) About how tall are those people who weigh 118 lbs?

The usual measure of size is the average, and the standard question here involves estimating the average of one variable for a subpopulation having a given value of the other variable.

Example D: What is the estimated average weight for people 65 inches tall?

The first thing to do is to decide which variable is the independent variable and which is the dependent variable. The independent variable is the one whose value is given—the dependent variable is the one whose average (or SD) the problem asks you to estimate. In Example D above, the independent variable is height (height in the subpopulation is given to be 65 inches) and the dependent variable is weight (you are asked to estimate the average weight for this subpopulation).

The procedure now involves converting the independent variable value given to SD's, multiplying by the correlation, then converting this result back to the units of the dependent variable.

The following program will lead you through this procedure:

1) Which is the independent variable? *height*

2) What is it's value? *65 inches*

3) How many SD's is it from its average? *-1* (since 65 inches is 3 inches below 68 inches, the average, and 3 inches $=$ 1 SD)

4) Multiply this last number by the correlation coefficient:
   -0.6 ($-1 \times 0.6 = -0.6$)

5) Multiply by the SD of the dependent variable:
   -9 lbs. (-0.6×15 lbs. $= -9$ lbs.)

6) Add the average of the dependent variable:
   111 lbs. (since 120 lbs. $+$ (-9 lbs.) $= 111$ lbs.)

Example E: A study gives the following data:

> average height of mother $=$ 65 inches, SD $=$ 2 inches
> average height of son $=$ 69 inches, SD $=$ 3 inches
> r $=$ 0.5

Estimate the average height of those sons whose mothers are 62 inches tall.

1) Which is the independent variable? _____

2) What is its value? _____

3) How many SD's is it from its average? _____

4) Multiply this last number by the correlation coefficient. _____

5) Multiply by the SD of the dependent variable. _____

6) Add the average of the dependent variable. _____

Computing the Regression Estimate

Solution: 1) mother's height; 2) 62 inches; 3) -1.5 (= (62-65)/2); 4) -0.75; 5) -2.25 inches (= -0.75 × 3 inches); 6) 66.75 inches (= 69 inches - 2.25 inches).

Problems

10.1 For the last data set, estimate the average height of those sons whose mothers are: a) 63 inches tall; b) 65 inches tall; c) 67 inches tall; d) 69 inches tall.

Estimate the average height of those mothers whose sons are: e) 63 inches tall; f) 66 inches tall; g) 69 inches tall; h) 72 inches tall.

10.2 A study gives the following results:

$$\text{average IQ} = 120; \quad \text{SD} = 10$$
$$\text{average LSAT} = 550; \quad \text{SD} = 100$$
$$r = 0.6$$

Estimate the average LSAT for those subjects with an IQ of:
a) 100; b) 110; c) 120; d) 125; e) 130; f) 135.

Estimate the average IQ of those subjects whose LSAT score is: g) 450; h) 500; i) 550; j) 600; k) 650; l) 700.

> m) For each 10 point increase in IQ, how much does the LSAT increase on average?
>
> n) Suppose one particular subject has an IQ of 100. Can you say what this person got on the LSAT?
>
> o) Consider the group of people with IQ = 100. Can you say exactly what their average LSAT score is?

10.3 Here are measurements of a sample of Berkeley meter maids:

$$\text{average height} = 66 \text{ inches}, \quad \text{SD} = 5 \text{ inches}$$
$$\text{average weight} = 140 \text{ lbs}, \quad \text{SD} = 40 \text{ lbs.}$$
$$\text{correlation coefficient between height and weight} = -0.90$$

a) In view of the negative correlation between height and weight for this group, you would say that this group looks: i) like a typical group of people from the general population; ii) somewhat more uniform than a group of typical people; iii) so similar that the members are almost indistinguishable from one another; iv) bizarre.

b) What is the average weight of those meter maids who are 6 feet 4 inches tall (76 inches)? i) 164 lbs; ii) 126 lbs; iii) 212 lbs; iv) 68 lbs.

c) What is the average weight of those meter maids who are 4 feet 8 inches tall (56 inches)? i) 164 lbs; ii) 126 lbs; iii) 212 lbs; iv) 68 lbs.

10.4 The pair of values: (average high school grade, average college grade) is recorded for a large number of UC Berkeley students chosen at random, The pair of values: (average high school grade, average college grade) is recorded for a large number of UC Berkeley students chosen at random, both grades being on a scale from 0 to 4. This yields a football-shaped scatter diagram. The correlation between the average high school and college grade is 0.7. The SD of the college average grades is 1.5 times the SD of the high school average grades. Two students have high school averages that are 0.6 points apart. How far apart would you expect their college averages to be?

10.5 Here are the summary statistics on a group of college graduates applying to law school:

ave. GPA = 3.3, SD GPA = 0.3, ave. LSAT = 600, SD LSAT = 50, r = .6

What is the regression estimate of the LSAT score of those students with GPA: a) 3.0   b) 3.3   c) 3.6

II. The Regression Fallacy

The regression fallacy is most vividly seen in test-retest situations. The idea is that for subjects with a given first score, the average second score is expected to be closer to average. Often this effect is interpreted as evidence of an effect that may not exist. To determine whether the regression effect is causing a difference, ask yourself:

1) What is the regression estimate of the average of the dependent variable?

2) Is the regression estimate about equal to the actual average of the dependent variable? If so, you are only observing the regression effect— if not, possibly something else is causing the difference.

Example A: The average on the LSAT is 500, the SD is 100. An LSAT prep school enrolls about 100 people who scored 300 their first time tested. On their second test, their average score was 350. The people who run the prep course claim that their course has added an average of 50 points to these people's scores. Altogether, 5000 people took both tests. The correlation between scores on the first test and scores on the second (for all 5000 people) was 0.6. Does the 350 average support the claim made by the prep course administrators? Discuss.

Solution: The regression estimate of the average score on the second test is 380 (= $0.6 \times (-2) \times 100 + 500$).

The Regression Fallacy

Problems

10.6 A group of pre-law students, all of whom got 550 on their first LSAT test, took a prep course to improve their performance. The overall average on each test was 500. The average score on the second test was 540. This is cited as evidence that the prep course did not help, and in fact hurt them. What does regression theory say?

10.7 A large group of pre-law students, all of whom got 550 on their first LSAT test, took a prep course to improve their performance. The average score on their second LSAT test was 550 (on each test, the overall average score for all students was 500, and the SD was 100). The fact that the average score of this group did not increase from 550 is cited as evidence that the prep course did not help. What does regression theory say?

10.8 If in problem 10.6, the average on each test is 500, the SD is 100 and the correlation is 0.9, what does the evidence suggest?

10.9 Suppose the correlation in the last problem is 0.6. Now what does regression theory say?

10.10 A group of pre-law students who averaged 520 on the LSAT take the prep course mentioned above to prepare for a second attempt. On the second examination, this group averages 540, while the overall average remains 500. Can this be explained by the regression effect, or does this indicate that the prep course helped? The correlation was 0.6.

10.11 One year in a large history course, the midterm scores averaged 60 with an SD of 20. The final scores also averaged 60, but the SD dropped to 15. Can this be explained by the regression effect?

10.12 A study is made of people who stutter. Each subject is asked to read two passages of equal length, and the number of times they stutter while reading each passage is recorded. The researchers discover that the subjects who stuttered many times on the first passage tended to stutter fewer times on the second passage. They conclude that the subjects who stuttered many times on the first passage must have been nervous the first time and more relaxed the second time, so that they tended to stutter less. What do you think?

10.13 An instructor standardizes his midterm and final so that the class average is 50 with an SD of 10 on both tests. The correlation between the tests is always around 0.5. On one occasion, he took all the students who scored below 30 at the midterm, and gave them special tutoring. They all scored above 50 on the final. Can this be explained by the regression effect?

III. Exercises

10.14 True/False

If the correlation coefficient is +1 then 100 % of the regression estimates are correct.

10.15 Given the following five points on a graph (with the first number x and the second y): (2,3) (1,5) (0,4) (3,6) (-1,2). Find the following: a) the average and SD of x; b) the average and SD of y; c) the correlation between x and y; d) draw an approximate picture, including the points, the SD line and the regression line, clearly labeled.

10.16 A group of children was given an IQ test when they were ten years old and again when they were 12. The scores are summarized here:

> average score at 10 = 100, SD = 20
> average score at 12 = 100, SD = 50
> r = 0.7

a) One child got a score of 120 in the 10-year-old test. Use the regression method to estimate his score on the 12-year-old test.

b) True/False: The regression effect says that this child is expected to score lower on the second test than on the first. Does this statement agree with your answer to a)? Explain.

10.17 A doctor records, for 400 hospital patients selected at random, the pair: (temperature when admitted, temperature 24 hours later). He finds that the average temperatures are 101 degrees when admitted, 99 the next day. The SD is 1.5 degrees for temperatures when admitted, 0.5 for those the next day. The correlation between the two temperatures is 0.8. The scatter diagram is football shaped.

a) If a patient enters with temperature 103, what do you expect his temperature to be the next day?

b) True/False: The fact that the average temperature the day after entering is 2 degrees less than upon admission is due to the regression effect.

Exercises

10.18 A coach records, for 100 athletes, their times in running the mile on two consecutive days. The summary statistics are:

> average day 1 = 4.5 minutes, SD = 0.5 minutes
> average day 2 = 5.0 minutes, SD = 0.5 minutes
> r = 0.8

a) If an athlete ran the mile in 5.0 minutes the first day, what would you expect her time to be the second day?

b) True/False: The fact that the average time on the first day is 0.5 minutes less than on the second day is due to the regression effect.

10.19 The regression line and the SD line are the same if: a) the SD of the dependent variable equals the SD of the independent variable; b) the correlation is 0; c) the correlation is 1; d) the average and SD of the dependent variable are the same as the average and SD of the independent variable.

10.20 A research team obtained the following measurements on 1000 subjects:

> average IQ = 100, SD = 15
> average daily caloric intake = 2000, SD = 200

The regression estimate of the IQ of subjects who consumed 2300 calories a day was 100. What can you conclude about the correlation coefficient?

10.21 Here is an incomplete list of points and corresponding scatter plot:

x: -1 0 0 ?
y: 0 0 1 ?

You are allowed to complete the list and scatter plot by choosing one more point from the following points: i) (1,1); ii) (1,-1); iii) (-1,1); iv) (-1,-1); v) cannot be done using i) through iv). Answer the following questions (hint: draw the scatter diagram):

a) Which option would you choose to make the correlation coefficient negative?

b) Which option would make the regression line (for estimating y from x) have a slope of 1/2 (i.e., a 45 degree line)?

IV. Solutions

10.1 a) 67.5 inches; b) 69 inches; c) 70.5 inches; d) 72 inches; e) 63 inches; f) 64 inches; g) 65 inches; h) 66 inches

10.2 a) 430; b) 490; c) 550; d) 580; e) 610; f) 640; g) 114; h) 117; i) 120; j) 123; k) 126; l) 129; m) 10 IQ points is 1 SD of IQ, which gives on average an increase of 0.6 SD's in LSAT score, that is 60 LSAT points. This also may be seen by looking at the difference between answers a) and b), or b) and c), or c) and e), or f) and d); in each of these pairs the regression estimates are for IQ's 10 points apart; n) No. 430 is only an estimate. This individual's score could be anything. 430 is the estimated average LSAT of all those with IQ = 100. If the study is large (in number of subjects), this estimate of the average is likely to be very close to the actual average LSAT score for those with IQ = 100. But even then, this particular person may be above or below average; o) No. Again, 430 is just an estimate of their average score.

10.3 a) iv; b) iv; c) iii

10.4 $0.7 \times 0.6 \times 1.5 = 0.63$

10.5 a) 600; b) 630 ($= 600 + 0.6 \times 1.0 \times 50$); c) 660 ($= 600 + 0.6 \times 2.0 \times 50$)

10.6 The information is inconclusive. Regression theory says that the average score of this group *should* decline, but you need the correlation coefficient to say how large a decline is expected.

10.7 Regression theory says the average should go down, towards 50— but it remained 550, so the prep course may explain the absence of a decline. That is, the evidence suggests that the prep course helped.

10.8 People with 550 on the first test will be on average about $0.5 \times 0.9 \times 100 = 45$ above average on the second. Thus you expect their average score on the second test to be 545. Their performance was lower. This suggests that the prep course hurt them.

Solutions

10.9 The expected average score on the second test is now $0.5 \times 0.6 \times 100 = 530$. The group did somewhat better than expected, and the data suggests the course was helpful.

10.10 The course might have helped. Due to the regression effect the score should have dropped to 512.

10.11 Yes. The lower scores rise a bit to the average and the higher scores decline nearer to the average.

10.12 No, the regression effect.

10.13 No, only some should go above 50.

10.15 a) average $= 1$, SD $= 1.414$; b) average $= 4$, SD $= 1.414$; c) $r = 0.7$; d) slope of SD line $= 1$, slope of regression line $= 0.7$

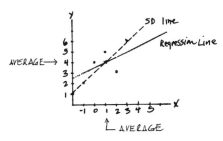

10.17 a) 99.53 ($= \frac{2}{1.5} \times 0.6 \times 0.5 + 99$); b) False—his is a reduction in the average of the whole population.

10.19 c)—in this case both have slope equal to $\frac{\text{SD of } y}{\text{SD of } x}$ and pass through the point of averages.

10.21 a) iv—(-1,-1); b) i—(1,1)

11

The RMS Error for Regression

I. Computing the RMS Error

The rms error of the regression line is the typical distance of the regression line from the points on the scatter diagram. In other words, it tells you how close the regression estimates are to actual values, on average. The rms error is easy to compute—it depends only upon the SD of the dependent variable and the correlation coefficient. The formula is:

$$\text{rms error} = (\text{SD of dependent variable}) \times \sqrt{1 - r^2}$$

Example A: average height = 68 inches, SD = 3 inches
average weight = 120 lbs, SD = 20 lbs.
r = 0.6

1) What is the rms error of the regression method for estimating weight from height?

Solution: Here the dependent variable is weight, so rms error is:

$$(20 \text{ lbs.}) \times \sqrt{1 - 0.6^2} = (20 \text{ lbs.}) \times .08 = 16 \text{ lbs.}$$

2) What is the rms error of the regression method for predicting height from weight?

Solution: The dependent variable is height, hence:

$$\text{rms error} = (3 \text{ inches}) \times \sqrt{1 - 0.6^2} = 3 \times 0.8 = 2.4 \text{ inches.}$$

Problems

11.1 Here is some IQ-GPA data for a college class:

average IQ = 120, SD = 10
average GPA = 3.1, SD = 0.4
r = 0.6

a) What is the rms error of the regression line estimating GPA from IQ?

b) What is the rms error of the regression line estimating IQ from GPA?

11.2 Researchers Tuddenham and Snyder obtained the following approximate weight figures for 66 boys:

> average weight (age 4) = 40 lbs, SD = 5 lbs.
> average weight (age 18) = 150 lbs, SD = 20 lbs.
> r = 0.4

Find the regression estimate for the weight of a boy at age 18 when his weight at age 4 was: a) 40 lbs; b) 45 lbs.

c) Find the rms error of the regression estimate for weight at age 18 from weight at age 4.

II. Visualizing the RMS Error

You should be able to look at a scatter diagram, draw in the approximate regression line, and estimate the rms error. You do this by looking at the vertical distances of the points from the line (the "errors") and then choosing a typical error—one that is not exceptionally large for the group but is also not exceptionally small. As a rough guideline, the point you choose should have an error larger than about 2/3 of the other points' errors and smaller than about 1/3 of the other points' errors.

Example A:

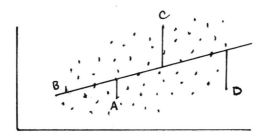

In the above scatter diagram, the approximate regression line is drawn and several "errors" made by the line are marked. One of these errors is about the size of the rms error of the line—which one?

Solution: Look at all the points on the scatter diagram and mentally note the size of each point's error—that is, its vertical distance from the line. Of the four points (A, B, C and D), point A comes closest to having a "typical" error—the errors of many points are larger and the errors of many others are smaller. So A's error is roughly equal to the rms error.

Problems

11.3 Below is a scatter diagram with its regression line. Several errors are marked. Which one is about equal to the rms error?

11.4 Here is a scatter diagram with its regression line drawn in, but without tick marks on the vertical scale. The rms error is 25. Beside it are four vertical scales—which one is the vertical scale of this scatter diagram?

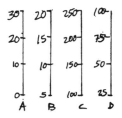
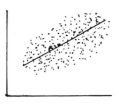

III. Computing the New Average and the New SD

In Chapter 10, you studied particular subpopulations—those with a given value of the independent variable. Using this value, you learned to estimate—for that subpopulation—the *average* of the dependent variable. In this section you will learn to estimate the *SD*. For ease of speaking, these estimates are called the "new average" and the "new SD".

Computing the New Average and the New SD

It is easy to compute the new SD: use the rms error. In formula form:

$$\text{new SD} = (\text{old SD}) \times \sqrt{1 - r^2}$$

Example A: average height = 68 inches, SD = 3 inches
average weight = 120 lbs., SD = 20 lbs
r = 0.6

1) What is the rms error of the regression estimate of weight from height?

$$(20 \ lbs.) \times \sqrt{1 - r^2} = 16 \ lbs.$$

2) Approximately what is the SD of weight for those people 68 inches tall?

New SD = rms error = 16 lbs.

3) Approximately what is the SD of weight for those people 72 inches tall?

The same as above, 16 lbs.

4) Approximately what is the SD of height for those people weighing 140 lbs?

New SD = (3 inches) × $\sqrt{1 - .6^2}$ = (3 inches) × 0.8 = 2.4 inches

Equipped with this formula and the method of the last chapter, you can now estimate both the average and the SD of one measurement for a subpopulation whose other measurement is given to you.

Problem

11.5 Given the above height-weight data, estimate the average and the SD of the weights for those who are: a) 65 inches tall; b) 68 inches tall; c) 62 inches tall.

Estimate the average and the SD of heights of all those who weigh:
d) 120 lbs; e) 130 lbs; f) 100 lbs.

IV. Approximating the New Average and New SD Geometrically

You should be able to estimate the new average and the new SD by looking at the scatter diagram. Here is a scatter diagram of data from a high
school GPA-IQ study:

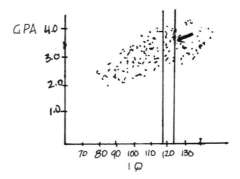

What is the average GPA of those with IQ = 120? Looking at the points
in the strip over IQ = 120, the average GPA seems to be about *3.4*—this
is the geometrical estimate. If you had the two averages, two SD's and
correlation you could use the regression (new average/new SD) method
to confirm this estimate—if the scatter diagram is truly football-shaped,
you should get about the same answer. Similarly, the SD of GPA in this
strip appears to be about *0.4*. The new SD computation should likewise
confirm this value. Of course you can also use the entire scatter diagram
to estimate the rms error—this will also give you an estimate of the new
SD.

Example A: Estimate the average GPA of those with IQ = 90.
Solution: The average GPA in the strip over 90 appears to be about 2.3.

V. Computing Percentages in Vertical Strips

Using the new average and new SD, you can also compute estimates of
percentages in vertical strips. Simply use the normal approximation, with
new average and new SD.

Example A: average height = 68 inches, SD = 3 inches
          average weight = 120 lbs, SD = 20 lbs.
          r = 0.6

In the subgroup 71 inches tall, estimate what percent weigh over 120 lbs.

Computing Percentages in Vertical Strips

Solution: The estimate of average weight for this subgroup (the new average) is 132 lbs, and the SD (new SD) is 16 lbs. Now draw a histogram of the subgroup:

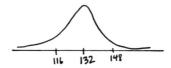

You want to find the area to the right of 120 lbs. This is a normal table problem. 120 lbs. is 12 lbs. below average on this histogram, or -0.75 SD's from average. The methods of Chapter 5 show that this area is about 50% + (55%)/2 = 77.5%.

Problems

11.6 Using the data:

> average height = 68 inches, SD = 3 inches
> average weight = 120 lbs, SD = 20 lbs.
> r = 0.6

a) What percent of those 65 inches tall weigh over 124 lbs?

b) What percent of those 68 inches tall weigh between 104 lbs. and 136 lbs?

c) What percent of those 62 inches tall weigh less than 80 lbs?

d) What percent of those weighing 120 lbs are 65.6 inches tall or taller?

11.7 As part of a large experimental course, a diagnostic quiz is given at the outset of the course. At the end of the course, the instructor collected the following statistics:

> Diagnostic quiz (20 points possible): average = 10, SD = 4
> Final exam (100 points possible): average = 60, SD = 15
> r = 0.6

Of those who scored six points on the diagnostic quiz, about what percent scored below average on the final?

VI. Approximating Percentages in Vertical Strips Geometrically

You should have a clear geometrical idea of what you are computing in the previous section. Here is a scatter diagram representing data from a height-weight study:

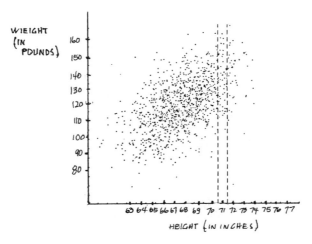

In the last section we looked at the subgroup with height 71 inches and estimated average weight in this subgroup to be 132 lbs. and the SD of weight to be 16 lbs. If you look at the vertical strip over 71 inches, you should be able to estimate the average and SD of weights in this strip. They are marked on the scatter diagram and do seem to be about 132 lbs. and 16 lbs. The percentage of this subgroup weighing over 120 lbs. is also marked. Roughly 3/4, or 75%, of the points in this strip have vertical measurement over 120 lbs.

Problems

11.8 Estimate geometrically the percentage weighing over 120 lbs, of those who are 65 inches tall.

11.9 Estimate geometrically the percentage weighing between 100 lbs. and 130 lbs, of those who are 68 inches tall.

11.10 Estimate geometrically the percentage weighing less than 130 lbs, of those 63.5 inches tall.

Exercises

11.11 True/False
A scatter diagram is homoscedastic. The farther from the average x is, the smaller the new SD(y) is.

11.12 A study obtained the following results for 1000 families:

average height of mother = 65 inches, SD = 2.5 inches
average height of son = 68 inches, SD = 3 inches
r = 0.6

a) Using the regression method, estimate the height of a son whose mother is 60 inches tall. b) Roughly what percentage of the sons were less than 68 inches tall? c) What is the rms error of the regression line for estimating the son's height from the mother's height? d) Of the sons who had mothers 60 inches tall, roughly what percentage were less than 68 inches tall?

11.13 This height-weight data collected from a large population:

average height = 68 inches, SD = 3 inches
average weight = 120 lbs, SD = 15 lbs.
r = 0.6

There are only five people 65 inches tall. Their weights are: 98 lbs, 106 lbs, 110 lbs, 114 lbs, and 122 lbs.

a) What is the average weight of this group of five? b) What is the SD of weight in this group of five? c) What is the regression estimate of weight from height for people 65 inches tall? d) Are a) and c) different? Why or why not? e) What is the error of the regression line which estimates weight from height? f) Are b) and e) different? Explain. g) Using regression, estimate the percentage of those people 65 inches tall who weigh more than 117 lbs. h) What is the *actual* percentage of this group weighing more than 117 lbs?

11.14 A study of a sample of 1000 families gave the following results:

average height of father = 68 inches, SD = 3 inches
average height of daughter = 63 inches, SD = 2.5 inches
r = 0.6

a) What is the rms error of the regression method for estimating daughter's height from father's height? b) Using the regression method, estimate the height of a daughter whose father is 62 inches tall. c) Roughly what percentage of the daughters were less than 63 inches tall? d) Of the daughters who had fathers 62 inches tall, roughly what percentage were less than 63 inches tall?

11.15 A large sample of school children was followed over time. One investigator looked at all the children who were at the 90th percentile in height at age 4. Some of these children turned out to be above the 90th percentile in height at age 18, while others were below. The number who were above the 90th percentile was:

a) quite a bit smaller than; b) about the same as; c) quite a bit larger than the number who were below. Is more information needed to determine this?

11.16 In a large study of the growth of children, the following results were obtained:

Girls age 13: average height = 58.5 inches, SD = 2.5 inches
Girls age 18: average height = 63.5 inches, SD = 2.5 inches

About what percentage of the 13-year-olds in the study were taller than 63.5 inches, the average for the 18-year-olds?

11.17 The heights and weights in a group are measured. In this group the SD of weight is 10 lbs. and the rms error of the regression line for estimating weight from height is also 10 lbs. The correlation coefficient between height and weight in this group is: a) 1.0; b) 0.60; c) 0; d) 0.40%; e) cannot tell from the data given.

11.18 Consider the regression line for estimating y from x in each of the following two lists:

List One

x:	1	1	2	2
y:	2	1	5	3

List Two

x:	2	2	4	4
y:	4	2	10	6

True/False: The rms error in List Two is the same as the rms error in List One. (Hint: you do not need to calculate.)

11.19 The heights and weights of a group of people are measured. The SD of weights in this group is 20 pounds. The rms error of the regression line for estimating weight from height for this group could be: a) 23 lbs; b) 12 lbs; c) 28 lbs; d) 30 lbs.

11.20 A study is done relating husband's age to wife's age. The regression line to estimate wife's age from husband's age is computed, as is the correlation coefficient. Which is the most plausible rms error of the regression line: a) 1 year; b) 4 years; c) 10 years; d) 15 years?

Exercises

11.21 The correlation between midterm score and final score for a Statistics class was 0.8. The average midterm score was 60 and the average final score was 70. The SD of midterm scores was 10 and the SD of the final scores was 15.

a) Suppose someone got 60 on the midterm. Estimate her score on the final. b) Consider those who got 60 on the midterm. The SD of their midterm scores was of course, zero. But they did not all get the same score on the final, so the SD of their final scores was not zero. What was it? c) What percentage of the group in b) scored above 79 on the final? d) About what was the average final score of all those who scored 80 on the midterm?

11.22 In a large study of fathers' heights and oldest sons' heights the following is obtained from 1000 families:

average father's height = 68 inches, SD = 3 inches
average son's height = 68 inches, SD = 3 inches
r = 0.6

Assume that the scatter plot is homoscedastic with both variables following the normal curve.

a) A father who is 63.5 inches tall has a percentile rank of about:
   i) 35%; ii) 68%; iii) 14%; iv) 7%; v) none of the above.

b) The percentile rank of a son is 69%. His height must be about:
   i) 71 inches; ii) 67 inches; iii) 69.5 inches; iv) 66.5 inches;
   v) none of the above.

c) One of the following is true. Which one?

   i) A father of above average height can expect his son to be 60% of his height; ii) A father of average height can expect his son to be below average due to the regression effect; iii) 60% of a son's height can be attributed to his father's height; iv) If in 5 years a father could add 1 inches to his height then he could expect his son to grow 0.6 inches taller in the same period; v) The taller a father is the taller (on the average) his son is likely to be.

d) Of those fathers who were 70 inches tall what percentage of their sons were above average (68 inches) in height?
   i) 50%; ii) 38%; iii) 69%; iv) 62%; v) none of the above

VIII. Solutions

11.1 a) 0.32 GPA points; b) 8 IQ points

11.2 a) 150 lbs; b) 158 lbs; c) rms error = 18 lbs.

11.3 c)—A is one of the largest errors made by the line, and B is one of the smallest. Only C) is most roughly "typical".

11.4 C) makes the rms error equal to about 25.

11.5 a) new average = 108 lbs, new SD = 16 lbs; b) new average = 120 lbs, new SD = 16 lbs; c) new average = 96 lbs, new SD = 16 lbs; d) new average = 68, new SD = 2.4; e) new average = 68.9 inches, new SD = 2.4 inches; f) new average = 66.2 inches, new SD = 2.4 inches.

11.6 a) 16%; b) 68%; c) 16%; d) 84%

11.7 New average (final) = 51, new SD (final) = 12. 60 is 0.75 SD's above the new average, so normal gives 77.5%.

11.8 About 10%

11.9 About 70%

11.10 About 100%

11.11 False—homoscedasticity means that the spread around the regression line is the same, regardless of x—it is equal to the rms error of the regression line.

11.13 a) 110 lbs; b) 8 lbs; c) 111 lbs; d) 110 is the actual average, 111 lbs. is only the regression estimate; e) 12 lbs; f) 8 lbs. is the actual SD; 12 lbs. is the rms error of the regression line, which only estimates the SD; g) new average = 111 lbs, new SD = 12 lbs, so 117 is 0.5 SD's above average. The answer, from the normal table, is about 31%; h) one out of 5, 20%

11.15 a)—the regression effect will cause the average height of the 4-year-old group to decline.

11.17 c)—because the rms error is equal to the SD of y only if r = 0.

11.19 b)—because the rms error can be at most equal to the SD of y.

11.21 a) 70; b) 9; c) 16%; d) 94.

12

The Regression Line

I. Computing Slope

The formula for the slope of the regression line uses the two SD's and the correlation coefficient: it is r multiplied by the ratio of the two SD's:

$$\text{slope} = r \times \frac{\text{SD (dependent variable)}}{\text{SD (independent variable)}}$$

The SD that goes on top is the SD of the variable the regression line is estimating.

Example A: In an income/education study, the SD of yearly income was $5,000 and the SD of education level (in years of schooling) was 4 years. The correlation was 0.3. What was the slope of the regression line for estimating: a) yearly income level from educational level? b) educational level from yearly income?

Solution: a) The dependent variable is yearly income—so the slope is: $0.3 \times \dfrac{\$5,000}{4 \text{ years}} = \$375$ per year; b) the dependent variable is educational level—so the slope is $0.3 \times \dfrac{4\text{-years}}{\$5,000} = 0.00024$ years per dollar. This can be expressed in years per $1,000: $0.3 \times \dfrac{4\text{-years}}{\$5,000} = 0.24$ years per $1,000.

Problem

12.1 In a study of age and income, the SD of age was 10 years, the SD of yearly income was $10,000, and r was 0.25. What was the slope of the regression line for estimating: a) yearly income from age? b) age from yearly income?

II. Interpreting the Slope

The slope of the regression line has an easy interpretation: it is the average increase in the dependent variable for a unit increase in the independent variable.

Example A: In a height-weight study of adults, the slope of the regression line for estimating weight from height is about 3 lbs./inch. Which of the following statements are true?

a) If two people differ by one inch in height, on average they differ by about 3 lbs. in weight.

b) If someone is 60 inches tall, their weight is, on average, $\frac{3 \text{ lbs.}}{\text{inch}} \times 60$ inches $= 180$ lbs.

c) An increase in height of 5 inches (from one person to another) is accompanied on average by an increase in weight of 15 lbs.

d) People have on average 3 lbs. of weight for every inch of height.

e) People have on average an additional 3 lbs. of weight for every additional inch of height.

Solution: a), c) and e) are correct, but b) and d) are not. In b), someone 60 inches tall has estimated weight (3 lbs./inch) $\times$ 60 inches + intercept—the 180 lbs. derived in b) 4 does not include the intercept, and therefore is incorrect. In d), likewise, the intercept is ignored. e) is, however correct—it is just a restatement of a).

Problem

12.2 A height-weight study of children is conducted. Assuming the slope of 3 lbs./inch in Example A is correct, would you expect the slope of the regression line for estimating weight from height for these children to be larger or smaller than 3 lbs./inch?

III. Computing the Intercept

The formula for the intercept of the regression line uses the slope and the two averages:

average (dependent variable) $-$ slope $\times$ [average (independent variable)]

Again, the dependent variable is the one the regression line is estimating.

How to Interpret the Intercept

Example A: In the income/education study in Example A of Computing Slope (above), the average income was $20,000 and the average educational level was 14 years. What is the intercept of the regression equation for estimating: a) yearly income from educational level? b) educational level from yearly income?

Solution:

a) average (yearly income) − slope × [average (educational level)

$$= \$20,000 - \frac{\$375}{\text{year}} \times 14 \text{ years} = \$20,000 - \$5,250 = \$14,750$$

Note that this can also be obtained using the regression method: the intercept is the regression estimate for a value of zero of the independent variable. An educational level of zero is 14 years below average, which is 3.5 SD's according to Example A. So the regression estimate of income is $3.5 \times 0.3 = 1.05$ SD's below average income. 1.05 SD's of income is $1.05 \times \$5,000 = \$5,250$, and $\$20,000 - \$5,250 = \$14,750$.

b) $14 \text{ years} - \dfrac{0.24 \text{ years}}{\$1,000} \times \$20,000 = 14 \text{ years} - 4.8 \text{ years} = 9.2 \text{ years}.$

This answer can be obtained using the regression method, as in part a).

IV. How to Interpret the Intercept

While the slope often has a physical meaning, the intercept often doesn't. It is often merely the number which makes the regression equation come out right and give the correct answers. However, if the independent variable can assume the value of zero, the intercept has the interpretation: the intercept is the regression estimate for the average value of the dependent variable when the independent variable is zero.

Example A: In the income/education example (Example A, above), the intercept of the regression line for estimating yearly income from educational level is $14,750. This has the interpretation that a person with zero years of schooling earns on average $14,750.

Example B: In the same example, the intercept of the regression line for estimating educational level from income is 9.2 years. This can be interpreted as: the average educational level of those in the study with no income is about 9.2 years.

Example C: In a height-weight study, the following data was computed:

average weight = 120 lbs., SD weight = 12 lbs.
average height = 67 inches, SD height = 2 inches
r = 0.6

The slope of the regression line for estimating weight from height is: $0.6 \times \dfrac{12 \text{ lbs.}}{2 \text{ inches}} = 3.6$ lbs./inch, and the intercept is:

$$120 \text{ lbs.} - \frac{3.6 \text{ lbs.}}{\text{inch}} \times 67 \text{ inches} = -121.2 \text{ lbs.}$$

How do you interpret the -121.2 lbs?

Solution: It is just the number that makes the regression equation work—no one weighs -121.2 lbs.

V. Exercises

12.3 True/False

a) If the SD of the dependent variable is the same as the SD of the independent variable, the slope of the regression line is equal to the correlation coefficient.

b) The correlation coefficient and the slope of the regression line are always of the same sign—either both are positive, both are negative, or both are zero.

c) The slope of the regression line is equal to r only if the two SD's are equal.

12.4 A study of college graduates is conducted relating the age at which they began college to the age at which they graduated. The regression line to estimate age at graduation from age at entrance is computed, as is the correlation coefficient. The slope of the regression line must be closest to: a) 4 years/year; b) 18 years/year; c) 1 year/year; d) 20 years/year.

12.5 A study is done relating husband's age to wife's age. The regression line to estimate wife's age from husband's age is computed, as is the correlation coefficient. The slope of the regression line must be closest to: a) 1 year/year; b) 0.5 years/year; c) 4 years/year; d) 8 years/year.

12.6 An investigator has data on 1,000 sets of SAT scores. He uses a computer program to calculate the regression line of math scores versus verbal scores, and to print out the residuals from the regression line. His assistant tries to do the same thing, but accidentally uses the residuals instead of the math scores. What is the slope and intercept of the assistant's regression line?

Exercises

12.7 For a class:

> average midterm score $= 70$, SD $= 5$
> average final score $= 55$, SD $= 20$
> $r = 0.5$

a) Find the regression equation

$$[\text{estimate} = \text{slope} \times (\text{x-value}) + \text{intercept}]$$

for estimating final scores from midterm scores.

b) A student scores 65 on the midterm; using a), estimate his final score.

12.8 An experimenter has a number of identical pieces of piano wire. A weight is attached to each and the length that the wire stretches is recorded. The weights used (in pounds) were 3, 5, 7, 9, . . ., 17, 19, 21. The correlation coefficient between weight and length of stretch is computed as $r = +0.99$, and the regression equation is:

$$\text{length of stretch} = +1.75 \text{ inches} + \frac{0.1 \text{ inches}}{\text{lb.}} \times \text{weight}$$

a) If a weight of 10 lbs. were put on a piece of wire like the ones used in the experiment, then we would expect: i) the wire would stretch very close to 2.75 inches; ii) the wire would stretch an amount only roughly around 2.75 inches; iii) can't tell because 10 lbs. was not a weight used in the experiment.

b) Consider the statement: "The slope of (0.1 inch)/lb. means that the wire stretches about 0.1 inches more if the weight is increased by 1 lb." The statement is: i) true for all weights; ii) true only for weights in the range from about 3 lbs. to about 21 lbs; iii) true only for weights which were used in the experiment.

12.9 A statistical analysis is made of midterm and final scores in a class.

> average midterm score $= 50$, SD $= 20$
> average final score $= 60$, SD $= 20$
> $r = 0.6$

The scatter diagram is football shaped.

a) Write the equation of the regression line for estimating final score from midterm score.

b) Of the students who scored 70 on the midterm, about what percentage scored over 88 on the final?

c) Of the students who scored 65 on the midterm, about what percent scored over 80 on the final?

12.10 A weight-losing clinic weighs people upon entering and one week later after the first session of the program. The scatter diagram of entering and first-week weights for the 400 people in the program is football shaped, with the following summary statistics:

> average entering weight = 190, SD = 15
> average first-week weight = 185, SD = 17
> r = 0.7

The histogram of entering weights follows the normal curve.

- a) Find an interval centered at the average first-week weight such that about 240 of the individuals have first-week weights in that interval.

- b) What is the equation of the regression line for predicting first-week weights from entering weights?

- c) If someone enters with a weight of 195 lbs, what would you estimate his first-week weight to be?

- d) Find an interval centered at the first-week weight determined in c) such that about 60% of the individuals with entering weights near 195 lbs. will have first-week weights in that interval.

- e) True/False: The fact that the average first-week weight is 5 lbs. less than the average entering weight is due to the regression effect. Why?

12.11 A GPA-LSAT study shows that the slope for estimating LSAT from GPA is about 100 LSAT points per GPA point. Which of the following are correct?

- a) If one person has twice the GPA of another, his or her estimated LSAT score is 100 points higher than the other's.

- b) If two people differ by 1.5 points in GPA, you can expect their LSAT scores to differ by about 150 points.

- c) A person with a GPA of 3.5 will score, on average, $3.5 \times 100 = 350$ on the LSAT.

- d) The SD of LSAT scores is $100 \times$ the SD of GPA scores in the group studied.

12.12 For the following data set, find the average and SD of x, the average and SD of y, and the correlation coefficient.

> x: −1   −1   −1   1   1   1
> y: −1   −1   1   −1   1   1

Solutions

VI. Solutions

12.1 a) $0.25 \times \dfrac{\$10,000}{10 \text{ years}} = \$250$ per year

b) $0.25 \times \dfrac{10 \text{ years}}{\$10,000} = 0.25$ years per $1,000

12.2 Smaller—an increase in height of one inch should be accompanied by an increase in weight somewhat smaller than among adults, since children are smaller in cross-section (and body volume).

12.3 a) True—this follows from the formula: slope $= r \times \dfrac{SD(y)}{SD(x)}$

b) True; c) True—this follows from the formula.

12.5 a—by elimination; 1 year/year is the only reasonable increase in age of husband for each year age of wife increases.

12.7 a) Slope $= 2$, intercept $= -85$; b) $2 \times 65 - 85 = 45$

12.9 a) Slope $= 0.6$, intercept $= 30$; b) 16%; c) 24.5%

12.11 Only b) is correct—a) is nonsense; c) ignores the intercept; and d) would be correct only if r were 1.0.

13

What Are the Chances?

We will use the following terms: experiment, trial, outcome, event. Here are their meanings.

Experiment: A series of operations done with a chance device.

Example A: Toss a coin 4 times.

Example B: Roll a die 4 times.

Example C: Toss a coin. If you get heads, roll a die 4 times. If you get tails, draw 2 cards from a shuffled deck. (This experiment involves several chance devices.)

Trial: One execution of an experiment.

In Example A, one trial consists of tossing the coin 4 times. If you did several trials of this experiment, you might get "H H T H" for instance, on the first trial, and "T H H T", on the second trial, etc.

Outcome: A complete description of the results of one trial of an experiment.

In Example A, "T H T H" is one possible outcome. But "at least one H" is not an outcome. If you were told only that "at least one H" occurred, you would not be able to say exactly what the result of the trial was. "At least one H" could mean any number of outcomes:

	H	T	T	T
or:	H	H	T	T
or:	H	H	H	T
or:	H	T	T	H
	.	.	.	.
	.	.	.	.
	.	.	.	.
		etc.		

In Example C, (heads, 6, 3, 3, 1) is an outcome; (tails, Ace of Spades, Jack of Diamonds) is another outcome.

Long Run

Event: Something that may or may not happen in an experiment.

In Example A, getting "at least one head" is an event.

In Example C, getting "no black cards" is an event.

Note that an event is a collection of outcomes. Also note that this language is slightly more specific than that used in FPP.

I. Long Run

These are the simplest probability problems. To solve them, ask yourself:

1) How many possibilities are there?

2) How many "qualify"?

3) Divide the number qualifying by the number of possibilities.

All of these experiments involve a number of outcomes that are equally likely to occur. This will not be the case with problems in later sections.

Example A: Roll a die once. What is the chance of getting a one-spot?

Solution: 1) How many possibilities are there? *Six, since the die has 6 faces.*

2) How many qualify? *Only one, since only one of the faces has a one-spot.*

3) So the answer is: *1/6.*

Example B: A box contains two red tickets and three blue tickets. A ticket is drawn from the box at random—what is the chance of "red"?

Solution: 1) How many possibilities are there? *Five—there are five tickets.*

2) How many qualify? *Two—two are red.*

3) So the answer is: *2/5.*

Problems

13.1 A lottery has 1000 tickets. You purchase five of them. One ticket is drawn out of the 1000 to determine the winner. What is your chance of winning?

13.2 A roulette table has 18 red slots, 18 black slots, and two green slots. What is the chance you will win if you bet on: a) red? b) black? c) green?

13.3 A box contains six red tickets and one green ticket. Five tickets are drawn at random without replacement. What is the chance the green ticket is among the five tickets drawn?

13.4 A box contains seven tickets numbered 1 through 7. Four tickets are drawn at random without replacement. If the ticket numbered "2" is among the four drawn, you win $1. What is the chance you win $1?

13.5 A deck of cards is shuffled. What is the chance the second card is a diamond?

II. Estimation of Chances

Sometimes you can estimate the chance of an event without actually calculating it. The two principles that are most useful are:

1) The chance that two events happen together is at most equal to the chance that either one happens individually.

2) The chance that either one or the other of the two events happens is at least equal to the chance that either one happens individually.

Example A: True/False. Deal two cards from a deck. The chance that "both are clubs" is less than 1/4.

Solution: True—the first card is a club 1/4 of the time, so certainly both cards are clubs even less often than that.

Example B: True/False. Toss a coin three times. The chance you get "at least one H" is greater than 1/2.

Solution: True—you get "H on the first toss" 1/2 of the time; but you get "at least one H" *at least* that often.

Problems

13.6 True/False

Two dice will be rolled.

a) The chance that "both fall aces" is smaller than 1/6.

b) The chance of getting "at least one ace" is more than 1/6.

13.7 A fair die is going to be rolled twice. The chance of getting "4" at least once is: a) 1/36; b) 1/6; c) larger than 1/6 but smaller than 1/3; d) 1/3; e) larger than 1/3.

Addition Rule

III. Incompatible Events

To decide if two events are incompatible, ask: "Can both events happen in the same trial of the experiment"? If the answer is "yes", the events are compatible. If the answer is "no", the events are incompatible.

Example A: Toss a coin four times.

Let Event A be: You get "at least three heads".
Let Event B be: You get "at least three tails".

Are A and B compatible?

Solution: A and B could happen separately in two different experiments—but could not happen in the same experiment, since if you get *at least* three heads you can have *at most* one tail. Therefore, A and B are *incompatible*.

Example B: Toss a coin four times.

Let C be the event you get "exactly two heads".
Let D be the event the "4th toss is tails".

Are C and D compatible?

Solution: Yes, they are compatible, since it's possible for both to happen. For instance, you could get the outcome "H T H T".

Problems

13.8 A coin is tossed five times. Which of the following pairs of events are incompatible?

a) "heads on the first toss" and "heads on the 5th toss"; b) "two heads on the first two tosses" and "three tails altogether"; c) "all the odd tosses are heads" and "three tails altogether"; d) "three heads on the first three tosses" and "tails on the last three tosses".

13.9 A five-card poker hand is dealt. Which of the following pairs of events are incompatible?

a) "all cards are red" and "the hand has three aces"; b) "at least two of the cards are red" and "all cards are lower than 10"; c) "all cards are hearts" and "there is at least one pair"; d) "at least four of the cards are hearts" and "there is at least one pair".

IV. Addition Rule

This is one of the basic rules used to solve probability problems. It says that the chance of an event composed of two or more incompatible events can be obtained by adding the chances of the incompatible events.

What Are the Chances? [Ch. 13]

The trick here is to re-express the event as one or the other of several incompatible events. This often involves a careful understanding of the language used to describe the event. Events that are incompatible must occur in different trials: if A happens in 20% of the trials, and B happens in a separate 30% of the trials, then one or the other (either A or B) will happen in 20% + 30% = 50% of the trials. The chance of an event measures how frequently it is expected to happen in a long series of trials.

Example A: A deck is shuffled. What is the chance that the Ace of Spades is one of the first two cards?

Solution: "Ace of Spades is one of the first two cards" can be re-expressed as two incompatible events:

> either A: "The first card is the Ace of Spades"
> or B: "The second card is the Ace of Spades"

Note that A and B are incompatible. The chance of A is 1/52, and the chance of B is 1/52, so chance of either A or B is 1/52 + 1/52 = 2/52.

Example B: A coin is tossed four times. The chance of getting "no H's" is 1/16, the chance of getting "one H" is 1/4, and the chance of getting "two H's" is 3/8 (the calculations behind these statements come in a later chapter). What is the chance of getting "two or fewer H's"?

Solution: "Two or fewer H's" is the same as either "no H's", "one H" or "two H's"; these events are all incompatible, so you can add the chances. Chance = 1/16 + 1/4 + 3/8 = 11/16.

Problems

13.10 Roll a die once. What's the chance of getting "an even number"?

13.11 Roll a die twice.

a) What is the chance that the "first roll is an even number"?

b) What is the chance that the "second roll is an even number"?

13.12 Shuffle a deck. What is the chance that the "first card is either an ace or red?"

13.13 True/False

Roll a die three times. The chance of getting "at least one 5" is 1/6 + 1/6 + 1/6.

13.14 True/False

A die is rolled four times. The chance of getting "at least one 6" is 4/6.

Independent Events

13.15 A box contains eight red tickets and one green ticket. Five tickets are drawn at random without replacement. What is the chance the green ticket is among the five tickets drawn?

V. Independent Events

Independence: Two events are independent if, when one happens, the chance that the other happens is unchanged.

Comment: Independence often arises in an experiment where the chance device is unchanged at each stage, i.e. drawing with replacement, and the events pertain to different stages of the experiment.

Example A: Toss a coin twice. "Heads on the first toss" and "tails on the second toss" are independent.

Example B: Draw 3 times with replacement from a box containing the numbers 1, 2, 3, and 4. "An even number on the first draw" and "the sum of draws two and three is odd" are independent events.

Problems

13.16 A box contains six colored balls with a number printed on each ball. The balls in the box have the following colors and numbers:

Red 1, Green 2, Red 2, Red 2, Green 1, Green 2

A ball is drawn at random from the box. Are color and number of the ball chosen independent? Explain.

13.17 Toss a coin twice. Decide which of the following pairs are independent.

a) "H on the first toss" and "H on the second toss"; b) "H on the first toss" and "T on the first toss"; c) "H on the first toss" and "at least one H"; d) "at least one H" and "at least one T"

13.18 True/False

A ticket is to be drawn at random from the box:

Red 1, Red 1, Red 2, Red 4, Blue 1, Blue 2, Blue 3, Blue 4

The color and number on the ticket drawn are independent.

13.19 A probability box: | 1 2 3 | has two items drawn from it *without* replacement. Are the outcomes on the two draws independent?

13.20 Toss a coin 100 times. Which of the following events are independent?

a) "at least 99 tosses come up heads" and "the first toss is a head";

b) "the first 50 tosses are exactly half heads and half tails" and "20 of the second 50 tosses are heads"; c) "there is at least one head" and "the last toss is a head"; d) "there is at least one head" and "there is at least one tail"

13.21 a) Are the two events "it rains in Portland, Oregon" and "it rains in Dayton, Ohio" independent?

b) Are the two events "it rains in Portland, Oregon" and "the stock market rises" independent?

VI. Multiplication Rule

The multiplication rule says that if two or more events are independent, the chance that they all will happen in the same trial can be obtained by multiplying their chances.

Example A: Toss a coin twice. What is the chance that you get "heads on the first toss" and "tails on the second toss"?
Solution: "H on the first" and "T on the second" are independent events. Each happens 1/2 of the time, so the chance that both happen in the same trial is $(1/2) \times (1/2) = 1/4$.

Example B: A die is rolled and a coin is tossed. What is the chance that "the coin turns up heads" and "the die shows a number greater than 2"?
Solution: $(1/2) \times (4/6) = 2/6$

Sometimes the trick is to re-express the event as a sequence of several events, as in the following example.

Example C: A coin is tossed four times. What is the chance of getting "all tails"?
Solution: "All tails" means "T on the first", "T on the second", "T on the third" and "T on the fourth", so the chance is $(1/2) \times (1/2) \times (1/2) \times (1/2) = 1/16$.

Problems

13.22 Two dice are rolled. What is the chance that "both fall ones"?

13.23 A die is rolled three times. What is the chance that "all three rolls show three or more spots"?

13.24 Roll three dice. What is the chance of getting "three one-spots"?

Multiplication Rule Without Replacement

13.25 True/False

Roll a die six times. The chance of: "neither a one nor a two" is $(4/6)^6$.

13.26 A number is drawn at random three times, with replacement, from the box:

$$| \ 1 \ 2 \ 3 \ 4 \ 5 \ |$$

What is the chance that "an odd number appears each time"?

13.27 The Dow Jones Industrial Index rises on about half the days of the year, independent of season. It rains in Portland, Oregon about 60% of the time. About what percent of the time does it happen that both the Dow Jones Index rises and it rains in Portland, Oregon?

VII. Multiplication Rule Without Replacement

Suppose you have a box containing three red tickets and three green tickets. Suppose you draw from the box twice, *without* replacement. What is the chance of getting two red tickets?

Of course drawing without replacement makes the draws dependent, so you might think that you can't use the multiplication rule. However, if you understand the logic behind this rule, you will still be able to use it. "Both draws are red" means "the first draw is red" and "the second draw is red". How often does this happen? "The first draw is red" happens 3/6 of the time. What part of *this 3/6 of the time* will you get "red on the second draw"? This 3/6 of the time, the box looks different on the second draw: it has three greens, but only two reds. So, this 3/6 of the time, the chance of getting "red on the second draw" = *2/5*. Therefore, we will get "red on both draws" *2/5 of 3/6 of the time*, or 6/30 of the time.

When you want to find chance using the multiplication rule without replacement, ask yourself:

1) What is the sequence of events needed to achieve the desired result?

2) What is the chance of the first event?

3) If the first event occurs, what does the box (deck, etc.) look like?

4) Using the new box (deck, etc.), what is the chance of the next event?

5) And so on—when you have all the chances, multiply them.

What Are the Chances? [Ch. 13]

Example A: Draw twice from a box with five red tickets and three green tickets, without replacement. What is the chance of getting "two green tickets"?

1) What is the sequence of events needed to achieve the desired result?

   "Green on the first draw", "green on the second draw"

2) What is the chance of the first event? *3/8*

3) If the first event occurs, what does the new box look like?

   R R R R R G G

4) Using the new box, what is the chance of the next event? *2/7*

Solution: The chance of getting two green tickets is $(3/8) \times (2/7) = 6/56$.

Example B: Draw three times from a box containing five red tickets and three green tickets, without replacement. What is the chance of getting three red tickets?

1) What is the sequence of events needed to achieve the desired result?

   R R R

2) What is the chance of the first event? *5/8*

3) If the first event occurs, what does the new box look like?

   R R R R G G G

4) What is the chance of the next event? *4/7*

5) If the second event occurs, what does the new box look like?

   R R R G G G

6) What is the chance of the next event? *3/6*

Solution: The chance of getting three red tickets is:

$(5/8) \times (4/7) \times (3/6) = 60/336$.

Problems

13.28 Shuffle a deck. What is the chance "the first two cards are red"?

1) What is the sequence of events needed to achieve the desired result?

_____

2) What is the chance of the first event? _____

3) If the first event occurs, what does the box (deck, etc.) look like?

_____

4) Using the new box (deck, etc.), what is the chance of the next event?

_____

Rule of Opposites

13.29 In two draws from a deck of cards, what is the chance that both cards drawn are spades?

1) What is the sequence of events needed to achieve the desired result?

_____

2) What is the chance of the first event? _____

3) If the first event occurs, what does the box (deck, etc.) look like?

_____

4) Using the new box (deck, etc.), what is the chance of the next event?

_____

13.30 Two decks of cards are on a table. The left deck is complete; the right deck is missing a spade. You draw a card from the left deck, then draw a card from the right deck. What is the chance you draw: a) A spade from the left deck? b) A spade from the right deck? c) A spade from both decks?

13.31 A deck is shuffled. A five-card poker hand is dealt. What is the chance that: a) The first two are aces; b) All are red; c) All are clubs.

VIII. Rule of Opposites

This rule says: the chance that an event happens can be obtained by computing the chance of the event *not* happening, and then subtracting this chance from one. Key words are "at least" and "at most"—these usually tip you off that this rule will solve the problem.

Example A: A die is thrown twice. What is the chance of getting "at least one 4"?

Solution: 1) The opposite of "at least one 4" is: *"no 4's at all"*

2) "No 4's at all" is the same as: *"no 4's on the first throw" and "no 4's on the second throw"*

3) The chance of this opposite event is $5/6 \times 5/6 = 25/36$, using the multiplication rule. Since the opposite happens 25/36 of the time, the event itself happens: *1 - 25/36 = 11/36 of the time*

Example B: A coin is tossed seven times. What is the chance of getting "heads at least once"?

Solution: 1) The opposite of "H at least once" is: *"no H's at all"*

2) "No H's at all" is the same as: *"all T's"*

3) "All T's" happens $(1/2) \times (1/2) \times \ldots \times (1/2) = (1/2)^7$ of the time. So the answer is:

$$1 - (1/2)^7$$

Problems

13.32 Two dice will be rolled. What is the chance of getting "at least one ace"?

13.33 A die is rolled twice. What's the chance of getting "at least one 6"?

13.34 A fair die is rolled four times. What is the chance of getting "a 1 or a 2 at least once"?

13.35 A box contains four red balls and six black balls. Two draws are made independently and with replacement. What is the probability that "at least one red ball" is drawn?

IX. Exercises

13.36 A deck of cards is shuffled and placed on a table. Before any cards are turned over, what is the chance that the first card drawn is a 9?

13.37 A deck of cards is shuffled. What is the chance that the first card drawn is a black 9?

13.38 A deck of cards is shuffled and placed on a table. Before any cards are turned over: a) What is the chance the top card is a heart? b) What is the chance the bottom card is a heart? c) What is the chance that either the top or bottom card is the Ace of Spades?

13.39 Assume that the chance of nuclear war in a given year is 5%; that this chance remains the same from one year to the next; and that each year is an independent trial. What is the chance that there will be no nuclear war in a 20-year period?

13.40 A box containing two "1's" and one "2" has two items drawn from it without replacement. What is the chance that the second item drawn is a "1"?

13.41 A deck of cards is shuffled and placed on a table. That is the chance that the Ace of Spades is among the top ten cards?

Exercises

13.42 Five friends hold five raffle tickets for a drawing in which there are 95 other tickets and one prize. What is the probability that one member of the group wins the prize?

13.43 Fill in the blanks, using one word from each pair below, to make up two true sentences.

"If two things are _____, and you want to find the chance that _____ will happen, you can _____ the chances".

  i) incompatible, independent
 ii) both, at least one
iii) add, multiply

13.44 A deck of cards is shuffled. What is the chance: a) the top card is the Ace of Spades? b) the 26th card is the Ace of Spades? c) the bottom card is the Ace of Spades?

13.45 In a shuffled deck, what is the chance: a) the first card is a heart? b) the 26th card is a heart? c) the second card is a heart?

13.46 A class consists of three freshmen, six sophomores, and 11 juniors. Three members are selected at random. Find the chance that: a) the second one chosen is a sophomore; b) the second is a sophomore and the third is a junior; c) the second and third are both sophomores.

13.47 Among the digits 1, 2, 3, 4, 5; first one number is chosen at random, then a second is chosen at random, from among the remaining four digits. Find the chance that an even digit is selected: a) the first time; b) the second time; c) both times.

13.48 True/False

A deck of cards is shuffled and placed on a table. The chance that both the top card and the bottom card are hearts is 1/16.

13.49 A box contains one black, one white, and two red balls. Three balls are drawn successively from the box. Find the probability that the third ball drawn is white if: a) each ball drawn is replaced; b) only the first ball drawn is replaced; c) only the second ball drawn is replaced; d) both the first and second balls are not replaced.

13.50 First draw a ticket from Box A, which contains two red tickets and a green ticket, then draw a ticket from Box B, which contains two green tickets and a red ticket. What is the chance that both tickets are green?

13.51 Two cards are picked randomly from a deck of cards. What is the chance that both are aces?

13.52 A box has three tickets: two reds and one green. Two draws are made. What is the chance that both colors are drawn: a) if the draws are made without replacement? b) if the draws are made with replacement?

13.53 The chance that all five cards of a poker hand (five cards dealt in the usual way from a regular deck of 52 cards) will be red is:—a) smaller than; b) larger than; c) the same as—the chance that all five tosses of a fair coin will show heads.

13.54 A survey finds that 60% of all drivers on the highway are wearing safety belts. Assume a box model in which the drivers in a two-car highway collision are chosen at random from a box in which 60% of the tickets say "safety belt" and 40% say "no safety belt".

a) Under this model, what is the chance that both drivers are wearing safety belts?

b) Can you criticize this box model? In reality, would you expect a higher or lower percentage of such collisions to occur?

13.55 I toss a coin three times. Then you toss it three times. What is the chance that both of us get at least one head?

13.56 Two draws are made at random with replacement from the box:

| 1 2 3 |

What is the chance of getting 1 at least once?

13.57 A die is rolled twice. What is the chance of getting "at least one ace"?

13.58 Roulette has 38 slots, of which 18 are red. You play roulette three times, each time betting on red. What is the chance that you win at least one of the bets?

13.59 Ten people are waiting to be interviewed for a job. Four of them are women. If the order in which they are interviewed is random, what is the probability that the first two selections are women?

13.60 A box contains four balls, two red and two white. In each of two unrelated drawings, a ball is drawn at random and replaced. Consider the events:

A: The ball drawn on the first draw is red.
B: The ball drawn on the second draw is white.

Exercises

a) Say whether each of the following is true, and explain: i) A and B are independent; ii) A and B are mutually exclusive.

b) Compute the probabilities of the following events: i) Both A and B happen; ii) Either A or B happens.

13.61 True/False

Two dice are rolled. The chance both come up ones is equal to $(1/6) \times (1/6) = 1/36$.

13.62 "If two events are {incompatible, independent} and you want to find the chance that {both, at least one} will happen, you can {add, multiply} the chances."

In the sentence above, which *two* sets of words create valid statements?

a) incompatible, both, add
b) incompatible, both, multiply
c) incompatible, at least one, add
d) incompatible, at least one, multiply
e) independent, both, add
f) independent, both, multiply
g) independent, at least one, add
h) independent, at least one, multiply

13.63 A drawer contains four red socks and six white socks. If two socks are chosen at random, without replacement, what is the probability of getting two red socks?

13.64 An astronaut's oxygen supply comes from two independent sources. Source A has probability 0.9 of working, and Source B has probability 0.8 of working. What is the probability that at least one of the sources is working?

13.65 In a certain city, if the weather bureau forecasts rain, they are right 80% of the time. If they forecast no rain, they are right 70% of the time. In a given period of 10 years, rain is forecast 20% of the days. What percentage of the days did rain fall?

13.66 Find the chance that three unrelated people were all born on different days of the week.

13.67 Find the chance that a poker hand contains at least one black card.

13.68 At the beginning of each week, one of the weekdays (Mon., Tue., Wed., Thur. or Fri.) is chosen at random for a surprise quiz. What is the chance that in the first three weeks the quiz is on Wednesday or later?

13.69 A die is rolled three times. What is the chance that all three rolls show three or more spots?

13.70 Two draws are made at random from a box containing the numbers 1, 2, and 3. What is the chance of getting a 1 at least once?

a) with replacement; b) without replacement.

13.71 A deck of cards is shuffled and placed on a table. What is the chance that both the top card and the bottom card are hearts?

13.72 An insurance company notices that in a certain territory only 30% of the automobiles are black, but that 50% of the two-car collisions involve at least one black automobile. Does this indicate that black automobiles are greater insurance risks? (Adopt a chance model in which the color of each car in a two car collision is chosen at random from an appropriate box.)

X. Solutions

13.1 5/1000

13.2 a) 18/38; b) 18/38; c) 2/38

13.3 5/7

13.4 4/7

13.5 The second card can be any one of the 52 cards in the deck—so there are 52 possibilities. Three of them qualify (i.e., are diamonds), so the answer is $13/52 = 1/4$. This problem commonly confuses beginners. You might be tempted to think that the chance the second card is a diamond depends on whether the *first card* is a diamond or not—and in fact it does, but no mention is made of the first card in this problem. The second card will be a diamond 1/4 of the time. Likewise, any other deck position—for instance, the 10th card—will be a diamond 1/4 of the time.

13.6 a) True—the first falls ace 1/6 of the time; but only part of this 1/6 of the time do both fall aces. b) True—you get "at least one ace" more often than you get an "ace on the first", which occurs 1/6 of the time.

13.7 c)—since "at least one 4" happens more often than "4 on the first roll" (1/6 of the time) and less often than $1/6 + 1/6 = 1/3$ of the time.

13.8 c) and d)

Solutions

13.9 a) and c)

13.10 "An even number" is the same as: either "2" or "4" or "6". These three events are incompatible, and each has chance 1/6. Chance = 1/6 + 1/6 + 1/6 = 1/2.

13.11 a) The chance is the same as in the previous problem, 1/2; b) 1/2

13.12 You might be tempted to break this down into two events: "ace" or "red". The chance of "ace" is 4/52; the chance of "red" is 26/52, so you would calculate the chance of this event as 4/52 + 26/52 = 30/52. This is wrong. "The first card is an ace" and "the first card is red" are not incompatible events (why?), therefore you cannot add the chances. A breakdown which works is: "the first card is an ace or red" is the same as either "the first card is a black ace" or "the first card is red". The chance of "black ace" is 2/52, the chance of "red" is 26/52, so the answer is 2/52 + 26/52 = 28/52.

13.13 False—this solution arises from the incorrect breakdown into "5 on the first" or "5 on the second" or "5 on the third". Each of these events has chance 1/6; but they are not incompatible, so their chances cannot be added.

13.14 False—just as in the previous problem. Note that this same reasoning could conclude that in six rolls the chance of getting "at least one 6" is 6/6 = 1, and such an event is clearly not certain to happen.

13.15 "Green ticket among the five tickets drawn" can be re-expressed as: either "first ticket is green" or "second ticket is green" or "third ticket is green" or "fourth ticket is green" or "fifth ticket is green"— so chance is: 1/9 + 1/9 + 1/9 + 1/9 + 1/9 = 5/9. Note: This problem is very similar to 13.3, and can be solved in the same way as 13.3.

13.16 Yes—if you know the color, the chance of any number is the same.

13.17 a) The coin has no memory. If you get "H on the first toss", the chance of getting "H on the second toss" is still 1/2. These are independent; b) If "H on the first toss" happens, "T on the first toss" cannot happen. So the chance of "T on the first toss" would be zero, not 1/2. These are not independent; c) If you get "H on the first toss", you automatically have "at least one H", so the chance of "at least one H" equals one. These are not independent; d) If you gets "at least one H", you have at most one T, so the chance of "at least one T" is certainly lowered. These are not independent.

13.18 False—if you know the color, the chances of the same numbers are different. For instance, if the ticket is red, "the chance of 3" is zero. Alternately, if you know the number, in certain cases you know the color—for instance, if the number is "3", the color must be blue.

13.19 No—for instance, if you know that a "2" was drawn the first time, the chance of "1 on the second draw" is 1/2 and not 2/3, as it would be if you knew nothing about the first draw.

13.20 a) If the first toss is a head, your chances of getting "99 heads" is slightly improved. Alternatively, if "99 of the tosses are heads", this means the first toss was probably a head. These are not independent; b) These events relate to separate parts of the experiment, so they are independent; c) If there is "at least one head", the chance that the "last toss is a head" is slightly increased. These are not independent; d) If the first ten are heads, we are likely to have many more than "1/2 heads" in this experiment. So the chance that we get "1/2 heads and 1/2 tails" is diminished. These are not independent. If you know that "at least one is a head", the chance of a tail on any given toss is slightly decreased. These are not independent.

13.21 a) No. If it rains in one city, it's probably winter, and the chance of rain in another city is therefore increased; b) Yes. There is no evidence for seasonality in the behavior of the stock market; these two events are probably quite independent.

13.22 $(1/6) \times (1/6) = 1/36$

13.23 $(4/6) \times (4/6) \times (4/6) = 64/216$

13.24 $1/216$

13.25 True

13.26 $(3/5)^3$

13.27 $(1/2) \times (60\%) = 30\%$

13.28 1) What is the sequence of events? *"first card is red", "second card is red"*

2) What is the chance? *26/52*

3) What does the new deck look like? *26 blacks, 25 reds*

4) What is the chance of the next event? *25/51*

The chance of the first two cards being red is: *(26/52)* × (25/51)

Solutions

13.29 1) What is the sequence of events? *"spade", "spade"*

2) What is the chance? *13/52*

3) What does the new deck look like? *12 spades, 13 hearts, 13 clubs, 13 diamonds*

4) What is the chance of the next event? *12/51*

The chance of both draws being spades is $(13/52) \times (12/51)$.

13.30 a) 13/52; b) 12/51; c) Since the draws from the separate decks are independent, the chance is $(13/52) \times (12/51)$.

13.31 a) $(4/52) \times (3/51)$
b) $(26/52) \times (25/51) \times (24/50) \times (23/49) \times (22/48)$
c) $(13/52) \times (12/51) \times (11/50) \times (10/49) \times (9/48)$

13.32 The opposite of "at least one ace" is "no aces". Chance of "no aces" $= (5/6) \times (5/6) = 25/36$, so the chance of "at least one ace" is: $1 - 25/36 = 11/36$.

13.33 $1 - (5/6)^2 = 11/36$

13.34 The opposite of "a 1 or a 2 at least once" is "no 1's or 2's". The chance of the opposite is: $(4/6)^4$, so chance is: $1 - (4/6)^4$

13.35 The opposite of "no reds" is "two blacks".
The chance of the opposite is: $(6/10) \times (6/10) = 36/100$.
The chance of getting "at least one red" is: $1 - 36/100 = 64/100$.

13.37 2/52, because there are two black 9's in the deck, and 52 cards total.

13.39 $(.95)^{20} = 36\%$

13.41 10/52

13.43 independent, both, multiply; incompatible, at least one, add

13.45 a) 1/4; b) 1/4; c) 1/4

13.47 a) 2/5; b) 2/5; c) $(2/5) \times (1/4) = 1/10$

13.49 a) 1/4; b) 1/4; c) 1/4; d) 1/4. Note that each ball has an equal chance of being drawn third—so the chance the white one is drawn third is 1/4, regardless of how the drawing is done.

13.51 $(4/52) \times (3/51)$

13.53 a) Since as you deplete the deck of red cards it gets progressively harder to draw another red card, whereas the coin gets another head just as easily as the first.

13.55 $(7/8)^2 = 49/64$

13.57 $1 - (5/6)^2$

13.59 $(4/10) \times (3/9) = 2/15$

13.61 True

13.63 $(4/10) \times (3/9) = 2/15$

13.65 $(80\%) \times (20\%) + (30\%) \times (80\%) = 16\% + 24\% = 40\%$

13.67 $1 - (26/52) \times (25/51) \times (24/50) \times (23/49) \times (22/48) = .97$

13.69 $(2/3)^3 = 8/27$

13.71 $(13/52) \times (12/51)$

14

More About Chance

I. Outcome Lists

The outcome list is a list of all possible outcomes, together with their chances, in an experiment. It is usually tedious to derive, but once you have it you can answer any question about chances in the experiment. If none of the three rules from the last chapter help you solve a problem, derive the outcome list.

Example A: A fair coin is tossed four times. What is the chance of getting at least two heads?

Solution: You might try to re-express "at least two heads" as one of several incompatible events. You might not succeed. Then you might try to re-express it as a sequence of independent events—but this will fail too. Then you might look at the opposite event: "at most one head". If you try to apply the addition or multiplication rules, you will probably still get stuck. Now you should resort to the outcome list.

$$
\begin{array}{ll}
\text{H H H H} & (1/2)(1/2)(1/2)(1/2) = 1/16 \\
\text{H H H T} & 1/16 \\
\text{H H T H} & 1/16 \\
\text{H T H H} & 1/16 \\
\text{T H H H} & 1/16 \\
\text{H H T T} & 1/16 \\
\text{H T H T} & 1/16 \\
\text{H T T H} & 1/16 \\
\text{T H H T} & 1/16 \\
\text{T H T H} & 1/16 \\
\text{T T H H} & 1/16 \\
\text{H T T T} & 1/16 \\
\text{T H T T} & 1/16 \\
\text{T T H T} & 1/16 \\
\text{T T T H} & 1/16 \\
\text{T T T T} & 1/16
\end{array}
$$

The above is a complete list of every possible outcome of this experiment, with its associated probability. Notice that the outcomes are listed in a simple order. First come those with four H's—there is only one: H H H H. Next are listed all those with exactly three H's—there are four of those: H H H T, H H T H, H T H H, and T H H H. Next come those with exactly two H's, etc. You might have trouble listing those with two H's—notice that these six outcomes are listed by groups: the first group consists of all those in which the first toss is H: H H T T, H T H T, and H T T H. The second group consists of all those in which the first time H appears is the second toss: T H H T and T H T H. The third group consists of all those in which the first time the H appears is in the third toss: T T H H. If you organize the list in a natural way, you will be certain to get all the outcomes, with no duplicates. The trick is to find a good way to organize the list.

Once you have the list, you can compute the chance of your event by checking off those outcomes which cause the event to occur and adding their chances. The outcomes corresponding to "at least two heads" are:

H H H H	H T H T
H H H T	H T T H
H H T H	T H H T
H T H H	T H T H
T H H H	T T H H
H H T T	

Each of these outcomes has chance 1/16, so the chance of this event is $11 \times 1/16 = 11/16$.

Note: It is not really necessary to derive *all* the outcomes. If you are careful, you can list only those outcomes corresponding to the event. If you choose this shortcut, just make certain that you do not omit any relevant outcomes.

Example A: A box contains one red ball and nine white balls. Two balls are drawn (without replacement). What is the probability of picking one red ball and one white ball?

Solution: R W: $(1/10) \times (9/9) = 1/10$
  W R: $(9/10) \times (1/9) = 1/10$

So the chance is $1/10 + 1/10 = 2/10$.

Outcome Lists

Example B: A fair coin is tossed four times. Find the chance of more heads than tails appearing.

Solution:

H H H H	1/16
H H H T	1/16
H H T H	1/16
H T H H	1/16
T H H H	1/16

So the chance is 5/16.

Problems

14.1 A fair coin is tossed four times. What is the chance of getting exactly two heads?

14.2 The toll on a bridge is 25 cents. In a coin box are kept eight coins: three quarters, two dimes and three nickels. The driver reaches into the box and draws out three coins at random. What is the chance the amount is less than enough to pay the toll?

14.3 The probability box: $\lfloor$ 1 2 3 $\rfloor$ has two items drawn from it *without* replacement. Using a list, show that the chance is 1/3 that the *second* item drawn is a 1.

14.4 In a given city the family size is distributed as follows:

number of children	%
0	40
1	30
2	20
3 or more	10

If two families are chosen at random from the population, what is the chance that they will have a *total* of two children between them (assume the city is large and thus the drawing is like drawing with replacement)?

14.5 The probability box: $\lfloor$ 1 1 2 $\rfloor$ has two items drawn from it *without* replacement. Using a list, show that the chance is 2/3 that the second item drawn is a 1.

14.6 Three random draws with replacement are made from $\lfloor 1\ 2\ 3\ 4 \rfloor$.

   a) Find the chance the sum of the draws is 3.

   b) Find the chance the sum of the draws is 12.

II. Exercises

14.7 A deck of cards is shuffled and placed on a table. What is the chance that at least one of the top two cards is a heart?

14.8 Smith has three children. So does Wesson. What is the chance that both have two boys and one girl?

14.9 A fair coin is tossed. If it falls heads, B pays A $1; if it falls tails, A pays B $1. They start with $2 each and continue playing until one of them is broke.

   a) What is the chance the game ends after exactly two tosses?

   b) What is the chance the game ends after exactly four tosses?

14.10 True/False

If an experiment has eight possible outcomes, and three are of a special kind, then the probability of a special outcome is 3/8.

14.11 A box contains three red and two blue balls. Four draws are made. What is the chance of getting exactly three red balls when drawing: a) with replacement? b) without replacement?

14.12 Two numbers are drawn randomly with replacement from the box: $\lfloor 1\ 2\ 2 \rfloor$. What is the chance that the sum of the two numbers drawn is equal to three?

14.13 On Friday, April 25, 1980, the San Francisco Chronicle reported a public opinion poll showing that 57% of the voters felt Proposition 10 allowed less rent control, while 43% felt it allowed more rent control. Among those feeling that Proposition 10 allowed less rent control, 60% opposed the measure. Among those feeling that it allowed more rent control, 50% were in favor of the measure and 50% were opposed. Based on this data, what percent of the public wanted more rent control?

14.14 Of those admitted to law school, 80% have scored 600 or more on the LSAT exam. Of those rejected by law school, 75% have scored less than 600. Of those scoring 600 or more on the LSAT, what percentage get admitted to law school?

   a) 80%; b) 75%; c) 20%; d) 25%; e) Cannot be determined from the data given.

Conditional Probability

III. Conditional Probability

Sometimes you are given partial information about the outcome of an experiment. This usually enables you to make a new computation of the probability of events in the experiment, conditional upon the information given. The idea is simple—just as in the simplest problems in this chapter. First you determine what the "whole" list of outcomes is. Then you determine which of these outcomes "qualify", or belong to the "part" list specified by the conditional information.

Example A: A coin is tossed three times. If you are told that at least one of the tosses was H, what is the chance exactly two of them were H?

Solution: If at least one toss was a H, then the possible outcomes are:

$$
\begin{array}{ll}
\text{H T T} & 1/8 \\
\text{T H T} & 1/8 \\
\text{T T H} & 1/8 \\
\text{H H T} & 1/8 \\
\text{H T H} & 1/8 \\
\text{T H H} & 1/8 \\
\text{H H H} & 1/8 \\
\end{array}
$$

The information "at least one of the tosses was H" tells us this event happens 7/8 of the time. What part of this 7/8 of the time do we get exactly two H's?

$$
\begin{array}{ll}
\text{H H T} & 1/8 \\
\text{H T H} & 1/8 \\
\text{T H H} & 1/8 \\
\end{array}
$$

Now H H T, H T H, or T H H happen 3/8 of the time. So the question is, what part of "7/8 of the time" is this "3/8 of the time"? The answer is: $(3/8)/(7/8) = 3/7$.

This is the *conditional chance* of exactly two H's, *given* at least one H.

A program for solving this kind of problem is:

1) List all outcomes corresponding to the "whole", or given condition. Add their chances—this sum is the chance of the "whole".

2) List all of those outcomes which "qualify," or belong to the "part". Add their chances—this sum is the chance of the "part".

3) Divide the "part" chance by the "whole" chance.

Problems

14.15 There are three boxes. One has two silver coins, another has one silver coin and one gold coin, and the third has two gold coins. One of the boxes is chosen at random, then a coin is selected at random from it. Suppose the coin turns out to be gold. What is the probability that the other coin in the box chosen is also gold?

14.16 Smith has two children. At least one of them is a boy. What is the probability that both of them are boys?

14.17 Wesson has two children. The older child is a girl. What is the probability that both children are girls?

14.18 Remington has two children. Exactly one of them is a girl. What is the chance that the older child is a girl?

14.19 True/False

If two events, A and B, are independent, the conditional probability that B happens, given that A does not happen, is the same as the conditional probability that B happens, given that A happens.

14.20 In a certain city, 80% of all licensed drivers have auto insurance. Those who have insurance have a 5% chance of being involved in an accident, while those who do not have insurance tend to be more cautious, and have an accident rate of only 3%. Suppose you are hit by a licensed driver. What is the probability that this driver has insurance?

14.21 A probability box $\lfloor 1\ 1\ 2 \rfloor$ has two items drawn from it without replacement. Find the chance that the second item drawn is a 1, given that the first item drawn is a 1.

4.22 In a particular community a fungus is attacking many of the corn crops. A third of the farms are on a hill and the remainder are in the valley. Three fourths of all the corn crops on the hill have been attacked by the fungus, while only 3/8 of those in the valley have been attacked. A farmer is selected at random.

  a) What is the probability that his crop has been attacked by the fungus?

  b) Given that the farmer's crop has been attacked by the fungus, what is the probability that he lives on the hill?

14.23 A coin is tossed ten times. If you are told that the first nine tosses are heads, what is the chance that the 10th toss is a head?

Solutions

14.24 A certain pregnancy test determines pregnancy with high probability. In particular, if a woman is pregnant the chance the test will give a positive reaction is 90%. If she isn't pregnant, the chance it will give a negative reaction is 85%. Suppose a woman takes the test and gets a negative result. The chance she is pregnant is: a) 85%; b) 90%; c) 15%; d) 10%; e) cannot be determined from data.

14.25 A die is tossed twice. Given that the sum is greater than 7, what is the chance of getting a pair (i.e., the same number on both tosses)?

14.26 You shoot at a target four times. Each time you have a 50% chance of hitting the target. What is the chance that you hit the target on your first shot, given that you hit it at least once?

14.27 A fair coin is tossed until the number of heads is more than the number of tails, or until four tosses have been made. Find: a) The probability that the coin is tossed four times; b) The probability that a head is followed by a tail at some point during the tossing; c) The probability of getting two or more heads given that at least one head occurred.

14.28 The parent generation of a certain species of guinea pig is 3/4 short-haired and 1/4 long-haired. The offspring of two long-haired parents is always long-haired. The offspring of a long-haired parent and a short-haired parent is long-haired 2/3 of the time and short-haired 1/3 of the time. The offspring of two short-haired parents are long-haired 1/3 of the time and short-haired 2/3 of the time. A long-haired guinea pig is chosen at random; what is the probability that both of its parents are short-haired?

IV. Solutions

14.1

H H T T:	1/16
H T H T:	1/16
H T T H:	1/16
T H H T:	1/16
T H T H:	1/16
T T H H:	1/16

So the chance is 6/16.

14.2

N N N:	$(3/8) \times (2/7) \times (1/6) = 1/56$
N N D:	$(3/8) \times (2/7) \times (2/6) = 2/56$
N D N:	$(3/8) \times (2/7) \times (2/6) = 2/56$
D N N:	$(2/8) \times (3/7) \times (2/6) = 2/56$

So the chance is 7/56.

14.3 The outcome list with probabilities is:

$$
\begin{array}{ll}
1, 2 & (1/3) \times (1/2) = 1/6 \\
1, 3 & 1/6 \\
2, 1 & 1/6 \\
2, 3 & 1/6 \\
3, 1 & 1/6 \\
3, 2 & 1/6
\end{array}
$$

The outcomes corresponding to "the second item drawn is a 1" are: 2, 1 and 3, 1. So the chance is $1/6 + 1/6 = 1/3$.

14.4

$$
\begin{array}{ll}
0, 2 & 0.4 \times 0.2 = 0.08 \\
1, 1 & 0.3 \times 0.3 = 0.09 \\
2, 0 & 0.2 \times 0.4 = 0.08
\end{array}
$$

So chance is $8\% + 9\% + 8\% = 25\%$.

14.5 Label the two 1's as 1A and 1B. Out of the six outcomes, four have the second item $= 1$.

$$
\begin{array}{ll}
1A, 1B & 1/6 \\
1B, 1A & 1/6 \\
2, 1A & 1/6 \\
2, 1B & 1/6
\end{array}
$$

So the chance is $1/6 + 1/6 + 1/6 + 1/6 = 4/6 = 2/3$.

14.6
a) 1, 1, 1, $(1/4)^3$
b) 4, 4, 4, $(1/4)^3$

14.7

$$
\begin{array}{lll}
\text{H} & \text{H} & (13/52) \times (12/51) \\
\text{H} & \text{not H} & (13/52) \times (39/51) \\
\text{not H} & \text{H} & (13/52) \times (39/51)
\end{array}
$$

14.9 a) H H: 1/4; T T: 1/4; So the chance is $1/4 + 1/4 = 1/2$.

b) H T H H: 1/16; H T T T: 1/16; T H H H: 1/16; T H T T: 1/16
So the chance is $1/16 + 1/16 + 1/16 + 1/16 = 1/4$.

Solutions

14.11 a)

outcome	probability
R R R B	$(3/5)^3 \times (2/5)$
R R B R	$(3/5)^3 \times (2/5)$
R B R R	$(3/5)^3 \times (2/5)$
B R R R	$(3/5)^3 \times (2/5)$

So chance is $4 \times (3/5)^3 \times (2/5) = 216/625$.

b) Since there are only five balls and four draws are made, "exactly three red balls drawn" is the same as "one blue left". The chance of "one blue left" is 2/5.

14.13 55.71%

14.15 Whole: SG, G: 1/6; GG, G: 1/3; Part: GG, G: 1/3
Part/Whole $= (1/3)/(1/3 + 1/6) = 2/3$

14.16 Whole: B B, B G, G B; Part: B B
Part/Whole $= (1/4)/(3/4) = 1/3$

14.17 Whole: B G: 1/4; G G: 1/4; Part: G G: 1/4
Part/Whole $= (1/4)/(1/2) = 1/2$

14.18 Whole: B G: 1/4; G B: 1/4; Part: B G: 1/4
Part/Whole $= (1/4)/(1/2) = 1/2$

14.19 True—since A and B are independent, B happens with the same chance whether A does or does not happen.

14.20 Whole: insured, accident: 0.8×0.05
               uninsured, accident: 0.2×0.03
     Part: insured, accident: 0.8×0.05

Part/Whole $= (0.8 \times 0.05)/[(0.8 \times 0.05) + (0.2 \times 0.03)] = 87\%$

14.21 50%

14.22 a) $(1/3) \times (3/4) + (2/3) \times (3/8) = 50\%$; b) 50%

14.23 1/2

14.24 e—it would be necessary to be able to compute the chance of "pregnant, negative" and "not pregnant, negative," but these values are not obtainable from the information given.

14.25 Whole: 2, 6; 3, 5; 3, 6; etc.
$(1 + 2 + 3 + 4 + 5 = 15$ outcomes)
Part: 4, 4; 5, 5; 6, 6

All of the outcomes are equally likely, so Part/Whole = 3/15.

14.26   Whole: $1 - (0.5)^4$;   Part: 0.5
Part/Whole $= 0.5/(1 - (0.5)^4)$

14.27

outcome	chance
H	0.5
T H H	$(0.5)^3$
T H T H	$(0.5)^4$
T H T T	$(0.5)^4$
T T H H	$(0.5)^4$
T T H T	$(0.5)^4$
T T T H	$(0.5)^4$
T T T T	$(0.5)^4$

a) (THTH) + (THTT) + (TTHH) + (TTHT) + (TTTH) + (TTTT) =
$6 \times (0.5)^4$
b) (THTH) + (THTT) + (TTHT) $= 3 \times (0.5)^4$
c) Whole: $1 - (TTTT) = 1 - (0.5)^4$
Part: T H H, T H T H, T T H H
Part/Whole $= (0.5^3 + 2 \times 0.5^4)/(1 - 0.5^4)$

14.28

	father	mother	child	chance
Whole	LH	LH	LH	$(1/4) \times (1/4) \times 1 = 1/16$
	LH	SH	LH	$(1/4) \times (3/4) \times (2/3) = 2/16$
	SH	LH	LH	$(3/4) \times (1/4) \times (2/3) = 2/16$
	SH	SH	LH	$(3/4) \times (3/4) \times (1/3) = 3/16$
Part	SH	SH	LH	$(3/4) \times (3/4) \times (1/3) = 3/16$

Part/Whole $= (3/16)/(8/16) = 3/8$

15

The Binomial Coefficients

I. Counting

The factorial part of the binomial formula counts the number of outcomes with a given number of "successes". You do this by asking yourself three questions:

1) How many trials (tosses, rolls, etc.) are there?
2) How many "successes" are desired?
3) How many "failures"?

Then assemble these numbers into the formula: $\dfrac{(\text{trials})!}{(\text{successes})!(\text{failures})!}$

Example A: A coin is tossed five times. How many outcomes are there with two heads and three tails?

1) How many trials are there? *5*
2) How many successes? *2*
3) How many failures? *3*

Solution: The answer is: 5!/2!3!

This further reduces to 10. There are 10 such outcomes. As a check, here is a list of the 10 outcomes:

```
 1 - H H T T T
 2 - H T H T T
 3 - H T T H T
 4 - H T T T T
 5 - T H H T T
 6 - T H T H T
 7 - T H T T H
 8 - T T H H T
 9 - T T H T H
10 - T T T H H
```

You can try to find more, but you won't succeed—there are exactly ten and they are all in this list. The formula tells you the correct number every time.

Problem

15.1 A coin is tossed seven times. How many outcomes are there with exactly three heads and four tails?

Example B: Roll a die seven times and count the number of aces. How many outcomes are there with three aces and four non-aces?

Solution: If you are only counting the aces there are just two possibilities on each roll—ace or non-ace. So you can use this formula to count the number of outcomes where "success" = ace and "failure" = non-ace. The answer is the same as in the previous problem: $7!/3!4! = 35$.

Notice in Example B that the number of outcomes is the same as in Example A even though the *chance* of each of these outcomes is different from the chance in the coin experiment—the chance of one of the coin outcomes is $(1/2)^7$, while the chance of one of the die outcomes is $(1/6)^3(5/6)^4$. This remark highlights the two parts of the binomial formula—the factorial part tells you how many outcomes there are and the chance part tells you the chance of each of these outcomes.

The factorial part can also be used to count other things—any list that can be expressed as words of fixed length consisting of two letters (like H and T), where one of the letters occurs a given number of times.

Example C: How many seven-letter words are there with two A's and five B's?

Solution: There are $7!/2!5!$ such words. This reduces to 21. You can check this answer by making the list:

$$A\ A\ B\ B\ B\ B\ B$$
$$A\ B\ A\ B\ B\ B\ B$$
$$\text{etc.}$$

If you make the list carefully, you will get 21 words of this type.

Example D: A class consists of 10 people. How many ways are there of selecting a committee of size three from the class?

Solution: The trick here is to imagine 10 people lined up:

$$*\quad *\quad *\quad *\quad *\quad *\quad *\quad *\quad *\quad *$$

Now you can consider any committee to form a ten-letter word, simply by assigning an A to the three on the committee and a B to those not on the committee—for instance, the committee consisting of the 1st,

How to Recognize a Binomial Experiment

3rd and 7th person forms the following ten-letter "word":

A	B	A	B	B	B	A	B	B	B
*	*	*	*	*	*	*	*	*	*

There are exactly as many such words as there are such committees, so you can count the number of committees by using the factorial formula The answer is: $10!/3!7! = 120$.

Problems

15.2 From a lot of 27 items, how many different ways may a sample of size 24 be chosen?

15.3 Consider an eight-person race in which we ignore the possibility of ties. If gold medals are given to the first three finishers, how many ways are there in which the medals can be awarded?

15.4 A waitress receives an order from a table for six mugs of beer (all the same brand) and three glasses of wine, but she has forgotten which person ordered what. Assume everyone at the table ordered.

a) How many different ways could the waitress distribute the drinks to the people at the table?

b) What is the chance that each person will get what she ordered?

II. How to Recognize a Binomial Experiment

To decide whether you can use the binomial formula, ask the following three questions about the experiment:

1) Are there two possibilities on each trial?

2) Is the chance of each the same each time?

3) Is there a fixed number of trials?

If the answer to all three questions is yes, then the experiment is a binomial experiment, and you can use the binomial formula.

Example A: Toss a coin four times.

1) Are there two possibilities on each trial? *Yes—H and T*

2) Is the chance of each the same each time? *Yes—1/2*

3) Is there a fixed number of trials? *Yes—4*

This is a binomial experiment.

Example B: A roulette wheel is spun 10 times. You count the number of reds.

1) Are there two possibilities on each trial? *Yes—red and not-red.*

2) Is the chance of each the same each time? *Yes—red has chance 18/38 on each trial and not-red has chance 20/38 on each trial.*

3) Is there a fixed number of trials? *Yes— 10.*

This is a binomial experiment.

Example C: Ten cards are dealt from a deck. The number of red cards is counted.

1) Are there two possibilities on each trial? *Yes—R and B.*

2) Is the chance of each the same each time? *No—the chance of red changes as the number of cards in the deck is reduced.*

This is *not* a binomial experiment.

Problems

15.5 A die is rolled five times. The number of times an ace appears is counted. Is this a binomial experiment?

15.6 Two dice are rolled twenty times. The number of times the sum is seven or less is counted. Is this a binomial experiment?

15.7 A coin is tossed until a head appears. Is this a binomial experiment?

15.8 A die is thrown ten times. The number of 1's, 2's, 3's, 4's, 5's and 6's occurring is recorded. Is this a binomial experiment?

III. Simple Binomial

Simple binomial problems are those where you are asked to find the chance of an exact number of heads, successes, etc. They require only one application of the binomial formula.

To solve a simple binomial problem, you must determine five quantities: number of trials, number of "successes", chance of "success" on one trial, number of "failures", and chance of "failure" on one trial. The following example makes clear how these five quantities are assembled into the formula.

Example A: A box contains five red balls, two green balls, and three yellow balls. Five draws are made with replacement. What is the chance that you get three greens?

Simple Binomial

Solution: Number of trials = 5
   Number of "successes" = 3
   Chance of "success" = 2/10
   Number of "failures" = 2
   Chance of "failure" = 8/10

$$\text{Chance} = \frac{5!}{3!2!}\left(\frac{2}{10}\right)^3\left[\frac{8}{10}\right]^2$$

Example B: A box contains six red tickets and a green ticket. Five tickets are drawn with replacement from the box. What is the chance of getting the green ticket exactly twice?
Solution: Number of trials = 5
   Number of "successes" = 2
   Chance of "success" = 1/7
   Number of "failures" = 3
   Chance of "failure" = 6/7

$$\text{Chance} = \frac{5!}{2!3!}\left(\frac{1}{7}\right)^2\left(\frac{6}{7}\right)^3$$

Example C: A fair coin is tossed six times. What is the chance of getting exactly two heads?
Solution: Number of trials = 6
   Number of "successes" = 2
   Chance of "success" = 1/2
   Number of "failures" = 4
   Chance of "failure" = 1/2

$$\text{Chance} = \frac{6!}{2!4!}\left(\frac{1}{2}\right)^2\left(\frac{1}{2}\right)^4$$

Problems

15.9 If you toss a coin ten times, what is the chance you get five heads and five tails?

15.10 A fair coin is tossed four times. What is the probability of getting exactly two heads?

15.11 A box contains six white balls and one red ball. Five draws are made with replacement. What is the chance of getting exactly one red ball?

15.12 A coin with $P(H) = 2/3$ is tossed four times. What is the probability of getting exactly two heads?

15.13 A fair coin is tossed six times. What is the chance of getting exactly two heads?

15.14 A box contains five white balls and one red ball. Four draws are made with replacement. What is the chance of getting exactly one red ball?

15.15 A computer program is designed to simulate drawing 400 times at random with replacement from the box:

$$\boxed{0\ \ 1\ \ 2\ \ 3\ \ 4\ \ 5\ \ 6\ \ 7\ \ 8\ \ 9}$$

Find the chance that the digit "7" appears exactly 44 times in the 400 numbers.

IV. Multiple Binomial

In multiple binomial problems, you must find the chance of getting something like: a certain number of successes or more, or less than a certain number of heads, etc. This merely involves applying the binomial formula to each of the values included in the specified range, and adding up.

Example A: A fair coin is tossed eight times. What is the chance of getting six or more heads?

Solution: "Six or more heads" means "either six, seven or eight heads":

$$\text{six heads: } (8!/6!2!) \times (1/2)^8 = 28/256$$
$$\text{seven heads: } (8!/7!1!) \times (1/2)^8 = 8/256$$
$$\text{eight heads: } (8!/8!0!) \times (1/2)^8 = 1/256$$

so the chance is $28/256 + 8/256 + 1/256 = 37/256$.

Problems

15.16 You answer all the questions on a ten-question True/False test by tossing a coin. What is the chance you get eight or more correct answers?

15.17 Someone answers all the questions on a 25-question True/False test by tossing a coin. Find the chance he gets more than 15 answers correct.

Sign Test

V. Sign Test

This is merely a multiple binomial calculation. It generally arises out of a randomized experiment with control group chosen by coin tosses, typified by the kangaroo experiment described in the Review Exercises to Chapter 15 of FPP.

Example A: A doctor testing the effect of a drug on blood pressure has 20 patients available, and randomly arranges them in pairs. In each pair one person is randomly chosen to get the drug. The other gets a placebo, a preparation looking and tasting like the drug, but having no chemical influence. In eight of the pairs the patient getting the drug experiences a larger decrease in blood pressure than the patient getting the placebo. Suppose in fact, that the drug has no effect, so that either person in a pair has the same chance of having the larger blood pressure decrease. If so, what is the chance that in eight or more of the pairs the patient with the drug will have a larger decrease?

Solution: If the drug has no effect, the chance that the non-placebo person in a given pair will have a larger decrease is $1/2$—so it's just like coin-tossing—and the question is: what is the chance that in ten coin tosses you will get eight or more heads? This is a multiple binomial problem, and the answer is:

$$(10!/8!2!)(1/2)^{10} + (10!/9!1!)(1/2)^{10} + (10!/10!0!)(1/2)^{10} = 56/1024$$

or about 5%.

Problems

15.18 In a now famous experiment [Weil, Zinburg, and Nelson, *Science*, Vol. 162, Dec. 13, 1968, pp. 1234-1242] the investigators studied the effect of smoking marijuana on nine adult subjects. To test effects on intellectual functioning, one part of the work dealt with each subject's scores in a test under two conditions: a) 15 minutes after smoking a placebo cigarette that had no marijuana, and b) 15 minutes after smoking a marijuana cigarette. The subjects were new to marijuana and did not know whether the drug was present in a cigarette. The order in which the two conditions were assigned was chosen at random for each subject.

It turned out that seven of the nine adults performed more poorly on the test after smoking the marijuana than they did after smoking the placebo. Suppose the marijuana had no effect, so that a subject would be just as likely to do more poorly on the test after the placebo as after the marijuana. In this case, what would be the chance of seven or more of the subjects performing more poorly after marijuana than after the placebo?

151

15.19 It is claimed that a vitamin supplement helps animals, rats in particular, to learn to run a maze. To test whether this is true, 20 rats are divided into pairs. In each pair, one rat is given a normal diet, and the other is treated with the vitamin supplement. At the beginning of the experiment a coin is tossed to decide which animal of each pair is to receive the normal diet. The rats are then timed as they learn to run the maze. In eight of the ten pairs, the treated rat learns the maze more quickly than the untreated one. If in fact the vitamin supplement has no effect, what is the probability that eight or more of the treated animals would learn the maze more quickly than their untreated partners?

15.20 In a short ESP experiment, a subject is asked to predict the outcome of a coin toss six times. The subject gets five out of six correct. Assuming the subject has no ESP powers at all, what is the chance that the subject would guess five or more correct out of six tosses?

15.21 Someone claims to be able to predict the toss of a coin. You decide to test his powers by tossing a coin eight times and counting the number of correct guesses. If in fact he is bluffing and his guesses are random, what is the chance that he gets six or more tosses correct?

VI. Exercises

15.22 a) A coin has $P(H) = 2/3$. It is tossed four times. What is the probability of getting exactly two heads?

  b) In the above experiment, what is the chance of getting at least one head?

15.23 The binomial formula is not suitable for a sequence of draws without replacement. Why not?

15.24 A box contains three red balls and two blue balls. Four draws are made. What is the chance of getting exactly three red balls when drawing: a) with replacement? b) without replacement?

15.25 True/False
Five cards are dealt from a shuffled deck. The chance of getting exactly three hearts is:
$$(5!/3!2!)(13/52)^3(39/52)^2.$$

15.26 Five cards are dealt from a shuffled deck. What is the chance of getting exactly three diamonds?
Note: this problem illustrates the fact that the counting formula of the binomial is useful in solving certain non-replacement problems. In the

Exercises

case of this problem, it is merely the probability part of the formula which does not apply.

15.27 True/False

If you toss a coin ten times, your chance of getting five heads and five tails is 1/2.

15.28 A quiz has ten True/False questions. A student will get an "A" if he scores eight correct answers. Two students do not study for the quiz and resort to pure guessing in each of the ten problems. What is the probability that both of them will get an "A"?

15.29 There are 20 men and 5 women in a class. Each day one person is chosen at random.
  a) What is the chance that a woman will be chosen on any day?
  b) What is the chance that a man is chosen on the first three days?
  c) What is the chance of choosing exactly two women in the first five days?
  d) What is the chance of choosing at least two women in the first five days?

15.30 A box contains five tickets numbered 1, 2, 3, 4, and 5. Three tickets are drawn at random without replacement. If the ticket bearing the number 2 is among the three, you win $1. What is your chance of winning $1?

15.31 A die is rolled four times. What is the chance that more than two spots show each time?

15.32 There are ten banks in a country. Each has a 2.0% chance of failing in a given year. The amount of money needed to rescue a failing bank would be $100 million. The government has an insurance fund of $200 million to cover such emergencies. However, if more than two banks fail in a given year, the government cannot pay the depositors, and a general economic panic and collapse will ensue.

  a) Assuming that banks fail independently of one another, what is the chance in a given year that such an event will occur?

  b) The assumption that banks fail independently of one another is in fact incorrect. If one bank goes bankrupt, this generally places a strain on the entire banking system, since this bank may in fact default on loans from other banks. So if one bank goes broke, the chance that any one of the others collapses would be in fact greater than 2%. Is the chance of a catastrophe underestimated or overestimated by the above calculation?

Assume the independent model stated above is valid, and assume that the chances are independent and do not change from year to year.

c) What is the chance of going five years without a catastrophe?

d) What is the chance of going 20 years without a catastrophe?

VII. Solutions

15.1 $7!/3!4! = 35$

15.2 $27!/3!24! = 2925$

15.3 $8!/5!3! = 56$

15.4 a) Think of the people lined up in a row. Placing the wines with three people and the beers with six people makes a nine-letter "word" with three W's (wines) and six B's (beers). The number of such words is $9!/3!6! = 84$. b) Clearly each of the possible ways is equally likely, and there are 84 ways, so the chance must be 1/84.

15.5 Yes

15.6 Yes

15.7 No. There are two possibilities on each trial, and their chances remain the same on each trial, but the number of trials is not fixed—it could be 1, 2, 3, or any other number, since you cannot say beforehand when the first head will appear.

15.8 No, since you are not counting only two possibilities on each trial.

15.9 $(10!/5!5!)(1/2)^{10} = 25\%$

15.10 $(4!/2!2!)(1/2)^4 = 6/16$

15.11 $(5!/1!4!)(1/7)(6/7)^4 = 39\%$

15.12 $(4!/2!2!)(2/3)^2(1/3)^2 = 30\%$

15.13 $(6!/2!4!)(1/2)^6 = 23\%$

15.14 $(4!/1!3!)(1/6)(5/6)^3 = 39\%$

15.15 $(400!/44!356!)(1/10)^{44}(9/10)^{356}$

15.16 $(10!/8!2!)(1/2)^{10} + (10!/9!1!)(1/2)^{10} + (10!/10!0!)(1/2)^{10} = 45/1024 + 10/1024 + 1/1024 = 56/1024$

Solutions

15.17 $(25!/16!9!)(1/2)^{25} + (25!/17!8!)(1/2)^{25} + \ldots +$
$(25!/25!0!)(1/2)^{25}$

15.18 $(9!/7!2!)(1/2)^9 + (9!/8!1!)(1/2)^9 + (1/2)^9 = 23/256$

15.19 $(10!/8!2!)(1/2)^{10} + (10!/9!1!)(1/2)^{10} + (1/2)^{10} = 56/1024$

15.20 $(6!/5!1!)(1/2)^6 + (6!/6!0!)(1/2)^6 = 7/64$

15.21 $(8!/6!2!)(1/2)^8 + (8!/7!1!)(1/2)^8 + (1/2)^8 = 14.5\%$

15.23 The binomial formula assumes that the chance of "success" is the same for all trials.

15.25 False. The chance of a heart is not the same on all five trials since the cards are drawn without replacement. Therefore, the binomial formula cannot be used.

15.27 False—it is $10!/5!5!(1/2)^{10} = 256/1024$, or about 25%.

15.29 a) $5/25$; b) $(20/25)^3$; c) $(5!/2!3!)(5/25)^2 \times (20/25)^3$;
d) $1 - [(20/25)^5 + 5 \times (5/25)(20/25)^4]$

15.31 $(4/6)^4$

155

16

The Law of Averages

I. Long Run/Short Run—The Law of Averages Simplified

One crude version of the "law of averages" is a simple principle: if you want the ordinary to happen, bet on the long run—if you want the extraordinary to happen, bet on the short run.

Example A: In tossing a fair coin, is it easier to get 64 or more heads in 100 tosses, or 640 or more heads in 1000 tosses, or are the chances the same?

Solution: Close to 50% heads is "ordinary"—64% or more is unusual. So your chances are better in the short run—it's easier to get it with 100 tosses.

Example B: A coin will be tossed either two times or 100 times. If the number of heads is exactly equal to the number of tails you will win $1. a) two tosses is better; b) 100 tosses is better; c) both are the same.

Solution: The answer is a). Although *close to* 50% heads is ordinary, *exactly* 50% heads is not—so choose the short run.

Example C: In tossing a fair coin, is it easier to get 60 or more heads in 100 tosses or 600 or more heads in 1000 tosses, or are the chances the same?

Solution: The answer is: 60 heads in 100 tosses.

Problems

16.1 Which is more likely to happen, five heads in ten tosses of a fair coin or 50 heads in 100 tosses of a fair coin?

16.2 Here are two situations:

a) A coin will be tossed 100 times. If it comes up heads 60 or more times you win $1.

Box Models

b) A coin will be tossed 1000 times. If it comes up heads 600 or more times you will win $1.

   i) Situation a) is better; ii) Situation b) is better; iii) Both offer the same chance of winning.

16.3 True/False

A coin is tossed an even number of times. The chance that the number of heads equals the number of tails goes up as the number of tosses increases, provided that the coin is fair.

16.4 Which of the following is more likely? Explain intuitively, without calculations.

   a) A fair coin is tossed 25 times and 10 or more of these tosses are heads.
   b) A fair coin is tossed 50 times and 20 or more of these tosses are heads.

16.5 True/False

The chance of getting between 45% and 55% heads in 100 tosses of a fair coin is greater than the chance of getting between 45% and 55% heads in 1000 tosses since in 1000 tosses there is more chance for variation.

II. Box Models

The trick in many problems is to start off with the correct box model. The standard problem in this chapter refers to repeated trials of the same experiment, and to a quantity which is computed as the sum of the values from these trials. These trials are modeled by random draws, with replacement, from a box. To decide what goes in the box, ask yourself:

a) What are the possible values on *one trial*?
b) What are the chances of these values?
c) How many trials are there?

The answer to a) tells what numbers go in the box. The answer to b) tells you how many of each number go in the box. The answer to c) tells you how many draws are made from the box.

Example A: A coin is tossed 50 times and the number of heads is counted. What is the box model?

   a) What are the possible values on one trial? On one trial, the number of heads can be *zero or one*. These are the only numbers that will appear in the box.

b) What are the chances of each value? "0" and "1" heads appear with equal chance—so they are in the box in equal proportions. *The box contains one "0" and one "1"* (Note: You could equally well put two "0's" and two "1's" in the box—the important thing is that "0" and "1" appear in equal numbers).

c) How many trials are there? *Fifty,* so the number of draws from the box is 50.

Example B: You bet $10 on red in 100 spins of the roulette wheel, computing how much you win after the 100 spins. What is the box model?

a) What are the possible values on one trial? On one trial you either win $10 or lose $10 (which is the same as winning −$10), so the only numbers appearing in the box are *+10 and −10.*

b) What are the chances of these values? A "+10" occurs when red comes up, which is 18/38 of the time. So *18/38 of the numbers in the box should be +10 and 20/38 of the numbers should be −10.* The easiest way to do this is to put 18 "+10's" and 20 "−10's" in the box. (Note: you could equally well put nine "+10's" and ten "−10's" in the box again; only the proportions are important).

c) How many trials are there? *100*

Example C: A multiple choice test has 100 questions. Each question has five answers. Suppose you guess at random on each question and your number of correct answers is counted. What is the box model?

a) What are the possible values on one trial? On one trial the number of correct guesses can be either *1 or 0. So the only numbers appearing in the box are 1 and 0.*

b) What are the chances of each value? *You have 1/5 chance of getting the correct answer, or a 1, and 4/5 chance of getting an incorrect answer, or a 0. So the box contains one "1" and four "0's".*

c) How many trials are there? *100*

Problems

16.6 A coin is tossed 100 times and the quantity, "the number of heads minus the number of tails" is computed. What is the appropriate box model?

16.7 A coin is tossed 100 times and the quantity, "twice the number of heads minus the number of tails" is computed. What is the appropriate box model?

Box Models

16.8 A coin is tossed 100 times and the quantity, "the number of heads plus the number of tails" is computed. What is the appropriate box model?

16.9 A roulette table has 38 slots, of which 18 are red. If you bet $1 on red and red comes up, you win $1—otherwise you lose $1. What is the box model for your winnings playing roulette 100 times, betting $1 each time?

16.10 Betting on a single number in Keno pays 2 to 1, and you have one chance in four of winning. Suppose you play 100 times, betting $1 on a single number each time. Fill in the blanks from the choices below.

Your net gain is like the (i)_____ of (ii)_____ draws made at random with replacement from the box (iii)_____. Explain.

   (i)   a) sum;   b) average;   c) percent
   (ii)   a) 1;   b) 2;   c) 4;   d) 10;   e) 100
   (iii) a) −$1, $2
          b) −$1, −$1, −$2
          c) −$1, −$1, −$1, −$2
          d) −$1, −$1, −$1, −$1, −$2
          e) −$1, −$1, −$1, +$2

16.11 A coin is tossed 500 times. After each 100 tosses, the total percentage of heads since the first toss is computed. Thus, if after 100 tosses 58 heads have come up, the percentage of heads in the first 100 tosses is 58%; if after 200 tosses 110 heads have come up, the percentage of heads is 55%; etc. In this way five numbers are collected: the percentage of heads in the first 100 tosses, the percentage of heads in the first 200 tosses, and so on up to the percentage of heads in the first 500 tosses. Here are four sequences of five percentages. Which one most closely resembles the sequence of five percentages you might expect in this experiment?

   a) 60%   42%   48%   56%   45%
   b) 55%   51%   49%   50%   51%
   c) 50%   50%   50%   50%   50%
   d) 45%   55%   46%   46%   56%

16.12 A penny is tossed 16 times and a nickel is tossed 1600 times. Which coin is more likely to come up heads:

   a) in exactly half of the tosses?
   b) in more than 60% of the tosses?

III. Solutions

16.1 Exactly 50% heads is extraordinary—the answer is the shorter run.

16.2 Sixty percent or more is extraordinary—the chance is better in the shorter run, a).

16.3 False—this is an extraordinary event and becomes less likely the more you toss.

16.4 b)—40% or more is ordinary; the chance of getting 40% or more is greater with the larger number of tosses.

16.5 False—between 45% and 55% is ordinary, so the chance is greater with the greater number of tosses.

16.6 a) On one trial (i.e. one toss), this quantity is either $+1-0=+1$ (when a head is tossed), or $0-1=-1$ (when a tail is tossed). Now $+1$ and -1 have equal chances, so the box model is 100 draws from the box $\lfloor +1 \;\; -1 \rfloor$.

16.7 On one toss, "twice the number of heads minus the number of tails" can be either $(2 \times 1) - 0 = 2$ or $(2 \times 0) - 1 = -1$. So the box model is 100 draws from the box $\lfloor +2 \;\; -1 \rfloor$.

16.8 On one trial, this quantity is $1 + 0 = +1$ (when heads is tossed), or $0 + 1 = +1$ (when tails is tossed), so the box model is 100 draws from the box $\lfloor +1 \rfloor$.

16.9 On one play you win either $1 or −$1. So the box contains only the numbers "+1" and "−1". One hundred draws are made. Here "+1" has an 18/38 chance of occurring, so the box contains 18/38 "+1's" and 20/38 "−1's". Hence 18 "+1's" and 20 "−1's" do the trick. One hundred draws are made.

16.10 (i) a: sum; (ii) e: 100; (iii) e: −$1, −$1, −$1, +$2

16.11 b—the longer the run, the closer the percentage of heads will be to 50%.

16.12 a) The penny. Exactly 50% heads is unusual, so the smaller the number of tosses the more likely we are to get exactly half heads. b) The penny. More than 60% heads is unusual, so the smaller the number of tosses the more likely we are to get more than 60% heads.

160

17

The Expected Value and Standard Error

I. Computing the Expected Value (EV)

In these problems you must first design a proper box model. After that, the rest is easy. To find the expected value of the sum of many draws, just average the numbers in the box and multiply by the number of draws.

Example A: You bet $1 on red 38 times in a row in roulette. What is the expected value of your winnings?

Solution: What is the box model? *38 draws from:* | 18 (+1) 20 (-1) |
What is the average of the box? *-2/38*
$EV = (-2/38) \times 38 = -2$

Example B: A multiple choice quiz has 20 questions. Each question has four choices. If you guess at random on all 20 questions, what is the expected number of correct answers on your quiz?

Solution: What is the box model? *20 draws from:* | 1 0 0 0 |
What is the average of the box? *1/4*
$EV = (1/4) \times 20 = 5$

17.1 A die is rolled 100 times, and the sum of the number of spots appearing on each roll is computed.

What is the box model? _____

What is the box average? _____

$EV = $ _____

17.2 Two hundred random draws are made with replacement from the following box.

| 1 2 3 4 5 6 |

The sum of the 200 draws is computed. What is the expected value of the sum?

161

17.3 One hundred random draws are made with replacement from the box:

$$\boxed{1 \quad 2 \quad 3 \quad 4 \quad 5}$$

and the sum of these 100 draws is computed. What is the expected value of the sum?

II. Computing the Standard Error

The standard error tells you how close to the expected value the actual observations will be. To compute the standard error of the sum of draws from a box, multiply the SD of the box by the square root of the number of draws from the box. It is helpful to your intuition if you then say to yourself: "The sum will be (EV) ± (SE)".

Example A: A box contains two tickets, one "+2" and one "+4". One hundred random draws are made with replacement, and the sum of the draws is computed.

a) What is the expected value of the sum?
b) What is the SE of the sum?

Solution: a) What is the box average? *3; EV = 3× 100 = 300*
  b) What is the box SD? *1; SE = 1 × $\sqrt{100}$ = 10*

So the sum will be around *300 ± 10*.

Example B: Four hundred draws are made with replacement from box:

$$\boxed{1 \quad 5 \quad 5 \quad 7 \quad 7 \quad 11}$$

a) What is the expected value of the sum of the draws?
b) What is the standard error of the sum of the draws?

Solution: a) What is the box average? *6; EV =6× 400 = 2400*
  b) What is the box SD? *3; SE = 3 × $\sqrt{400}$ = 60*

Since EV = 2400 and the SE = 60, the sum will be about: *2400 ± 60*.

Problems

17.4 We make 25 random draws with replacement from the box:

$$\boxed{1 \quad 3 \quad 4 \quad 5 \quad 7}$$

and compute the sum of the numbers drawn. Find the EV and the SE of this sum.

Box average = _____

EV = _____

Box SD = _____

SE = _____

So the sum will be about _____ ± _____

Interpreting the SE

17.5 We make 225 random draws with replacement from the box:

$$\boxed{0 \quad 4 \quad 6 \quad 8 \quad 12 \quad -1}$$

and the sum of the numbers drawn is computed. What are the EV and the SE of this sum?

17.6 You play roulette 64 times, betting $1 on red each time. What are the EV and SE of your winnings?

III. Interpreting the SE

If you draw a number of times from a box at random and with replacement and compute the sum of the draws, the value that you get will be about the expected value, plus or minus the standard error. If you perform this experiment many times and record the sum each time, what will the list of successive sums look like? In particular, about how large will they be on the average, and how spread out will they be? The answer is that their average will be close to the EV and their SD will be close to the SE. Moreover, their histogram will be approximately normal.

Example A: A coin is tossed 100 times and the number of heads is counted. This 100-toss experiment is repeated many times and the number of heads in each experiment is recorded. Which of the following lists most plausibly represents the sequence of values obtained?

a) 49, 51, 50, 51, 48, 50, 51, 49, . . .
b) 41, 48, 60, 65, 35, 50, 40, . . .
c) 48, 55, 52, 43, 56, 52, 49, . . .

Solution: The EV of the number of heads is 50 and the SE is 5. So the sequence of observed values should average about 50 and have an SD of about 5. All three sequences have averages of about 50, but "a" has an SD much smaller than 5, and "b" has an SD much larger than 5—only "c" has both the right average and the right SD.

Problems

17.7 A box contains two tickets, one labeled "-1" and one labeled "+1". One hundred draws are made with replacement, and the sum of the draws is computed. This same procedure is repeated four more times, so that five numbers are obtained, each being the sum of 100 draws from the box. Below are four sequences. One is a plausible representation of what the five sums might look like. Which one is it? Why?

	1st Sum	2nd Sum	3rd Sum	4th Sum	5th Sum
a)	-9	+4	+11	-2	-1
b)	-21	+26	-19	-17	+18
c)	-1	-3	+2	+1	0
d)	+15	+11	+17	+14	+12

17.8 The expected value for a sum is 45, with an SE of 5. The chance process generating the sum is repeated ten times. Which of the following is the sequence of observed values?

a) 40, 45, 50, 40, 45, 50, 40, 45, 50, 40

b) 46, 52, 43, 47, 52, 56, 53, 36, 48, 43

c) 46, 44, 48, 47, 43, 42, 48, 45, 44, 42

17.9 Consider the sum of 25 draws with replacement from the box:
| 0 2 3 4 6 |.
The chance experiment for determining the sum is repeated a number of times. One of the following is the sequence of observed values. Which one?

a) 76, 84, 95, 85, 100, 70, 91, 103, 68, 97

b) 75, 76, 74, 76, 77, 73, 75, 72, 74, 76

c) 50, 100, 97, 104, 58, 71, 83, 112, 49, 65

d) 84, 70, 91, 80, 74, 50, 65, 82, 88, 81

IV. Normal Curve

The normal approximation methods used in Chapter 5 carry over directly to chance calculations when you have a number of draws from a box with a replacement. To calculate the chance the sum will fall in a given range draw the normal curve using EV in place of the average and SE in place of the SD—the area over the interval is the chance you want.

Example A: A box has average = 5 and SD = 2. A hundred draws are made with replacement. What is the chance the sum of these draws is between 490 and 510?

Solution: EV = $5 \times 100 = 500$, and the SE = $2 \times \sqrt{100} = 20$; so the sum will be about 500 ± 20. The normal curve looks like:

490 500 510

Normal Curve

and the shaded area represents the chance of getting a sum between 490 and 510. 490 is 1/2 SE below the EV and 510 is 1/2 SE above the EV, so the endpoints of the interval translate to -0.5 and +0.5. The area between -0.5 and +0.5 under the normal curve is 38%—this is the chance of getting a sum between 490 and 510.

Example B: You play roulette 100 times, betting $1 on red each time. What is the chance you come out ahead?

Solution: The box is: | 18 (+1)   20 (-1) |; the box average is -2/38; the box SD =1; EV = (-2/38) × 100 = -$5.25; SE = 1 × $\sqrt{100}$ = $10. So you expect to lose $5.25 ±$10. The normal curve is:

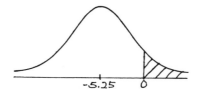

and the shaded region represents the chance that you come out ahead (win zero or better). Zero translates to 0.525 in standard units, so the area to the right of zero is the area under the normal to the right of 0.525. From the table, this is about 30%. You have about a 30% chance of coming out ahead in 100 plays of this game.

Problems

17.10 You play roulette 100 times, now betting $10 on red each time. Does this improve your chances of coming out ahead?

17.11 You play roulette 900 times, betting $1 on red each time. What is the chance you come out ahead?

17.12 You play the same game, except 10,000 times. Now what is your chance of coming out ahead?

17.13 What is the chance of getting between 40 and 60 heads in 100 tosses of a coin?

17.14 What is the chance of getting between 40% and 60% heads in 100 tosses of a coin?

17.15 What is the chance of getting between 775 heads and 825 heads in 1600 tosses of a coin?

V. Computing the SD of a 0-1 Box

A 0-1 box is a box containing only 0's and 1's. Its average and SD are computed from the list of 0's and 1's in it. Such a list, consisting entirely of 0's and 1's, will have an average between 0 and 1, and an SD which is smaller than 1. In fact, we will see that the SD can be at most 0.5. The average and SD of such a list can be computed using the methods explained in Chapter 4. But a shortcut formula exists that reduces the computational work. The average is easy—it is just the fraction of 1's in the list. The formula for the SD is: $\sqrt{(\text{fraction of 1's}) \times (\text{fraction of 0's})}$.

Example A: Compute the average and SD of the list: 0, 1, 1, 1, 1; first using the methods of Chapter 4 and then using the above formulas.
Solution: Using the methods of Chapter 4,
the average is: $(0 + 1 + 1 + 1 + 1)/5 = 4/5 = 0.8$.

The SD^2 is: $\dfrac{(.8^2 + .2^2 \, .2^2 \, .2^2 \, .2^2)}{5} = \dfrac{(.64 + 4 \times .04)}{5} = \dfrac{0.8}{5} = 0.16$.

So SD is: $\sqrt{0.16} = 0.4$.
Using the new formula: average $= 4/5$; SD $= \sqrt{4/5 \times 1/5} = 0.4$.

Problem

17.16 Compute the average and SD of the following $0-1$ lists. First use the procedures of Chapter 4 and then use the "shortcut" formulas.

a) 1, 1, 1, 0, 0
b) 1, 0, 0, 0, 0, 0, 0, 0, 0, 0
c) 0, 0, 0, 0, 1
d) 1, 1, 0, 0, 0

VI. Exercises

17.17 A person agrees to pay you $10 if you get a "one" or a "two" at least once in four rolls of a fair die. How much would you agree to pay him when you lose, to make the game fair?

17.18 The expected value of the number of red cards in a poker hand is: a) smaller than; b) larger than; c) the same as: the expected value of the number of heads in 5 tosses of a fair coin.

17.19 A coin is tossed 100 times and the sum "number of heads plus number of tails" is computed.

a) What is the expected value of this sum?
b) What is the standard error of this sum?

Exercises

17.20 True/False: The SE cannot be larger than the expected value.

17.21 Boxes A, B, and C each contain ten tickets. The average and the SD of the numbers in each box are given below.

Box A	Box B	Box C
average = 3	average = 3	average = 2
SD = 1	SD = 3	SD = 11

Sixteen draws are made at random with replacement from box A and the sum of the numbers drawn is calculated. This is also done for Box B and for Box C. In scrambled order the sums for the three boxes turn out to be 62, 84, and 50. Which of the following matchings is most plausible?

a)	62 - A	84 - B	50 - C
b)	62 - A	84 - C	50 - B
c)	62 - B	84 - A	50 - C
d)	62 - B	84 - C	50 - A
e)	62 - C	84 - A	50 - B
f)	62 - C	84 - B	50 - A

Give your reason.

17.22 True/False: As the number of draws increases the SE of the sum gets smaller and smaller.

17.23 In one trial of an experiment, a nickel and dime, both fair coins, are tossed together. This is repeated 100 times. On each of the 100 trials, you win $1 only if both coins come up heads; otherwise you win nothing.

a) What is the appropriate box model?
b) What is your chance of winning on the first trial?
c) What is the expected total amount you would win on the 100 trials?
d) What is the SE of the total amount you win?

17.24 A simplified roulette wheel has two red slots, two black slots, and one green slot. Suppose a $1 bet on green gives you winnings of $4 (plus your dollar back) if green appears.

a) Draw the box model for the winnings on a $1 bet on green.
b) This game is: i) favorable; ii) unfavorable; iii) fair.

You play 400 times, betting $1 on green each time.

c) What is the expected value of your winnings?
d) What is the SE of your winnings?
e) Approximately what is the chance you win more than $40?

17.25 Three random draws with replacement are made from the box:

$$\boxed{0 \;\; 0 \;\; 0 \;\; 1 \;\; 2}$$

a) What is the chance that no "0's" are drawn?

One hundred draws are made with replacement from the same box.

b) What is the chance that "1" turns up on 24 or more draws?

c) What is the chance that the sum of the 100 draws is greater than 72?

17.26 You play roulette 100 times, staking $1 each time. There are two possibilities.

I. You bet on a split each time. II. You bet on a number each time. Which of the following statements are true?

a) The chance of coming out ahead is the same for I as for II.

b) The chance of winning $10 is better with II than with I.

c) The chance of losing $10 is higher with II than with I.

17.27 A carpenter is measuring the distance between five points A, B, C, D, and E. They are all on a straight line. He finds that each of the four distances AB, BC, CD, and DE measure one yard, give or take an inch or so. These measurements are made independently, by the same procedure The distance from A to E is four yards, give or take around: a) 4 inches; b) 2 inches; c) 1 inch; d) 1/2 inch; e) 1/4 inch.

17.28 Which of the following games would you choose if given a choice?

Game A: Toss two dice and receive in dollars the sum of points showing

Game B: Toss a coin whose sides are marked "1" and "2" and receive in dollars twice the square of the number appearing.

17.29 A true/false exam consists of 25 questions. Each correct answer carries five points. However, to discourage guessing 2 points are subtracted for a wrong answer. If you do not attempt a question, you score zero for that question. The total possible score thus ranges from -50 to +125. The exam is given to a class of 100 students. Suppose a student decides not to attempt 9 questions and to guess at random (with a 50% chance of being right) on the remaining 16 questions.

a) What is his/her minimum possible score (Assuming that he made no errors on the questions he knew the answers for)? b) What is his/her maximum possible score? c) Show an appropriate box model. d) What is his/her expected score? e) What is the SE of the score?

Solutions

Suppose that each student in the class guesses randomly (with a 50% chance of being right) on all 25 questions.

f) Show the appropriate box model. g) Write down the expected value and SE of an individual's score under the model.

17.30 What is the chance of getting between 4900 and 5100 heads in 10,000 tosses of a coin?

17.31 What is the chance of getting between 49% and 51% heads in 10,000 tosses?

17.32 A cantaloupe farmer produces cantaloupes that average 12.0 ounces in weight, with an SD of 1.0 ounce. The cantaloupes are packed 16 to a box.

a) What is the chance that a given box has more than 12.25 lbs. of cantaloupes?

b) A shipment consists of 100 boxes. What is the chance it contains less than 1199 lbs?

VII. Solutions

17.1 Box model: One hundred draws from the box: $\lfloor 1\ 2\ 3\ 4\ 5\ 6 \rfloor$; box average = 3.5; EV = $3.5 \times 100 = 350$

17.2 Average of box = 3.5; EV = $3.5 \times 200 = 700$

17.3 Average of box = 3.0; EV = $3.0 \times 100 = 300$

17.4 The box average = 4; EV = $4 \times 25 = 100$; Box SD = 2; SE = $2 \times \sqrt{25} = 10$. The sum will be 100 ± 10.

17.5 Box average = 6; box SD = 4; EV = $6 \times 225 = 1350$; SE = $4 \times \sqrt{225} = 60$

17.6 Box = $\lfloor 18\ (+1)\quad 20\ (-1) \rfloor$; box average = -2/38; box SD = 1; EV = $(-2/38) \times 64 = -3.37$; SE = $1 \times \sqrt{64} = 8$. You expect to lose $3.37 plus or minus $8.

17.7 a) EV = 0, SE = 10

17.8 b) —Average is about right as is the SD.

17.9 The box has average = 3 and SD = 2. The number of draws is evidently 25 in each case, but only in case d) is the SD of the sample right: about $\sqrt{25} \times 2 = 10$.

17.10 The box is: $\lfloor$ 18 10 20 -10 $\rfloor$; The box average = -20/38; SD = 10; EV = (-20/38) × 100 = -$52.50; SE = 10 × $\sqrt{100}$. The normal curve is:

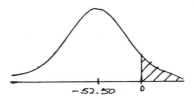

zero translates to 0.525 SE's above the EV, so the chance is again 30%. Increasing the bet does not change the chance that you come out ahead—it only magnifies the amount you win or lose.

17.11 Box = $\lfloor$ 18 (+1) 20 (-1) $\rfloor$; box average = -2/38; box SD = 1; EV = (-2/38) × 900 = -$47.25; SE = 1 × $\sqrt{900}$ = 30; so you expect to lose $47.25 ± $30. The normal curve is:

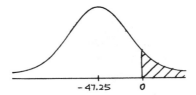

Zero is about $47.25 above the EV, which is 1.575 standard units (47.25/30 = 1.575). The area to the right of 1.575 under the normal curve is about 6%—this is your chance of coming out ahead.

17.12 EV = (-2/38) × $10,000 = -$525; SE = 1 × $\sqrt{10,000}$ = 100; so you will lose $525 plus or minus $100. The chance that you will get zero or more is the area to the right of zero, and zero translates to 5.25 SE's above the EV. The area to the right of 5.25 under the normal curve is far less than 1%—effectively, you have no chance at all of coming out ahead in 10,000 plays.

17.13 Box: $\lfloor$ 0 1 $\rfloor$; average = 0.5; SD = 0.5; EV = 0.5 × 100 = 50; SE = 0.5 × $\sqrt{100}$ = 5; so the number of heads will be around 50 plus or minus five. In standard units 40 converts to -2 and 60 converts to +2, and the area is A(2) = 95%. There is about a 95% chance of getting between 40 and 60 heads.

170

Solutions

17.14 Forty percent heads means 40 heads—sixty percent heads means 60 heads. So this is the same as the previous question, and the answer is the same—95%.

17.15 Box: $\boxed{0\ \ 1}$; average = 0.5; SD = 0.5; EV = 0.5×1600 = 800; SE = $0.5 \times \sqrt{1600}$ = 20; In standard units 775 translates to -1.25 and 825 translates to +1.25. A(1.25) = 79%

17.16 a) Average = 0.6, SD = $\sqrt{\frac{3}{5} \times \frac{2}{5}}$ = 0.49; b) average = 0.1, SD = 0.3; c) average = 0.2, SD = 0.4; d) average = 0.4, SD = 0.49

17.17 $(.8)10 = .2x$; x = \$40

17.19 a) (number of heads) + (number of tails) = 100 automatically, so the expected value is 100; b) zero

17.21 A: EV = 48, SE = 4; B: EV = 48, SE = 12; C: EV = 36, SE = 44. Here 84 could realistically only come from C. Of the two remaining boxes only B could realistically produce 62. So 84 is from C, 62 is from B, and 50 is from A. The answer is d).

17.23 a) $\boxed{1\ 0\ 0\ 0}$; b) 1/4; c) \$25; d) $\sqrt{100} \times \sqrt{(1/4)(3/4)}$ = \$4.3

17.25 a) 27/125; b) about 16%; c) average = 0.6, SD = 0.8, EV = 60, SE = 8, chance = 6.5%

17.27 b

17.29 a) -32; b) 80; c) $\boxed{+5\ -2}$; d) 1.5×16 = 24; e) SD = 3.5, SE = 3.5×4 = 14; f) same as c); g) EV = 37.5, SE = 3.5×5 = 17.5

17.31 This is the same as the previous problem—the answer is the same, 95%.

171

18

The Normal Approximation for Probability Histograms

I. The Continuity Correction

To make the normal approximation even more exact, you can use the so-called "continuity correction". This is the technique introduced on page 289 of FPP, and it involves moving the endpoints of the region up or down half a step. The rule here is always to *enlarge* the intervals to the nearest half integer at each included endpoint.

Example A: In one hundred tosses of a coin, find the chance of getting more than 45 heads.

Solution: Enlarging this interval by 0.5 leads to the interval "45.5 or more". The EV = 50 and the SE = 5, so "45.5 or more" converts to "−0.9 or more" under the normal curve. Therefore the chance is $A(-0.9) + (1/2)(1 - A(0.9))$.

Example B: Convert "55 or more" to its proper half-integer interval.

Solution: This interval has only one endpoint. Enlarging it by 0.5 at this endpoint gives "54.5 or more".

Example C: Convert "between 42 and 57 inclusive" to its proper half-integer interval.

Solution: This interval runs from 42 at the left to 57 at the right. Enlarging it by 0.5 at each endpoint gives "from 41.5 to 57.5".

Example D: Convert "more than 42 but less than 57" to its proper half-integer interval.

Solution: This interval runs from 43 on the left to 56 on the right. Enlarging it by 0.5 at each endpoint gives "from 42.5 to 56.5".

Normal Approximation, Ranges

Problems

18.1 Convert the following integer intervals to their corresponding half-integer intervals.

a) more than 25; b) greater than 10; c) larger than 100; d) at least 25; e) greater than or equal to 10; f) larger than or equal to 100.

18.2 Convert the following integer intervals to their corresponding half-integer intervals.

a) less than 25; b) less than 10; c) smaller than 100; d) at most 25; e) less than or equal to 10; f) smaller than or equal to 100.

18.3 Convert the following integer intervals to their corresponding half-integer intervals.

a) between 40 and 60 inclusive; b) between 40 and 60 exclusive; c) more than 100 but less than 200; d) at least 100 but at most 200; e) at least 100 but less than 200; f) more than 100 but at most 200.

II. Normal Approximation, Ranges

Problems

18.4 A coin is tossed 100 times and the difference:
(number of heads) − (number of tails) is computed.

a) What is the expected value of this difference?

b) What is the standard error of this difference?

c) Estimate the chance that the difference is between −5 and 5, inclusive.

18.5 A fair coin is tossed 10,000 times. Estimate the chance of getting:
a) 5050 or more heads; b) 5500 or more heads.

18.6 A die will be rolled 1800 times. Estimate the chance that there will be 300 or more ones.

18.7 One hundred draws are made at random with replacement from the box: $\boxed{0\ 0\ 0\ 1\ 2}$. What is the chance that "1" turns up on exactly 20 draws?

18.8 Suppose that in one week at a certain casino, there are 100,000 independent plays at roulette. On each play, the gamblers stake $1 on the number 17. The chance that the casino will win more than $5000 from these 100,000 bets is closest to:

a) 25%; b) 50%; c) 75%.

III. Normal Approximation, Exact Numbers

Problems

18.9 A computer is programmed to simulate drawing 400 times at random with replacement from the box: $\boxed{0\ 1\ 2\ 3\ 4\ 5\ 6\ 7\ 8\ 9}$.

a) Write down an expression for the exact chance that the digit 7 appears exactly 44 times in the 400 numbers.

b) Using the normal approximation compute this chance approximately.

18.10 Three draws are made at random with replacement from the box: $\boxed{1\ 2\ 3\ 4}$. The average of the numbers in the box is 2.5 with an SD of approximately 1.3.

a) Use the normal approximation to find the chance that the sum of the draws is between 4 and 11 inclusive.

b) Find (exactly) the chance that the sum of the draws is between 4 and 11 inclusive.

18.11 In 100 coin tosses, what is the chance of getting exactly 50 heads?

IV. Exercises

18.12 A box contains ten tickets, five marked with a positive number and five with a negative number. All the numbers are between −10 and 10. We will make 500 random draws with replacement from the box.

a) In order to estimate the chance that the sum will be positive, you need to know: (i) nothing else; (ii) the number of SD's in the box and the number outside; (iii) the average and the SD of the numbers in the box; (iv) the square root of the number of draws times the SD of the box; (v) the number of draws times the square root of the box; (vi) all the numbers in the box.

b) Repeat this to estimate the chance of getting 230 or more positive numbers.

c) Repeat to estimate chance of getting exactly 100 draws equal to "3".

18.13 A box contains ten numbers: $\boxed{1\ 1\ 1\ 1\ 2\ 2\ 2\ 2\ 3\ 3\ 4}$. One hundred draws are made from the box at random with replacement. What is the chance that in 100 draws:

a) the sum of the draws is less than 210?

b) the number "3" is drawn more than 24 times?

c) the average of the draws is between 1.85 and 2.1?

Exercises

18.14 a) Three draws are made with replacement from the box: $\boxed{1\ 0\ 0\ 0\ 0}$.

What is the chance that no "0's" are drawn?

A hundred random draws with replacement are made from the same box.

b) What will be their approximate: (i) average; (ii) SD?

c) What is the chance that "1" turns up on 24 or more draws?

18.15 A box contains five tickets: $\boxed{2\ 2\ 2\ 2\ 7}$.

a) Nine draws are made with replacement. What is the chance that seven or more "2's" are drawn?

b) Suppose 900 draws are made with replacement. What is the chance that the number of even numbers drawn is 700 or more?

18.16 A class is given a true/false quiz with ten questions. The teacher wants to discourage guessing, so wrong answers will be given negative points. A correct answer is worth $+1$ point and an incorrect answer is worth -2 points. What is the expected total score for a student who:

a) knows the right answer to six questions and tosses a fair coin on the remaining four? b) knows the right answer to five questions, tosses a fair coin on two others and doesn't know the remaining three? c) If you are sure of the answers to four questions, uncertain about four others, and have no idea about the remaining two, what is your best strategy if "uncertain about" means you have probability 3/4 of guessing the question right and "no idea" means you have probability 1/5 of guessing the question right?

18.17 Playing roulette, you bet $1 on red 400 times. Estimate the chance that you come out ahead.

18.18 A town contains 6000 families, each with three children. The chance that not more than 2325 of these families have two girls and a boy is: a) too small to be computed; b) about 98%; c) about 2325/6000, or about 39%; d) about 2%.

18.19 2500 draws are made at random with replacement from the box: $\boxed{1\ 3\ 4\ 5\ 7}$. The SD of the box is 2.

a) Approximately what is the chance that the sum of the draws is greater than 9900?

b) Approximately what is the chance that the number of "1's" drawn is between 500 and 530?

18.20 A penny is tossed 16 times. Approximately what is the chance that the penny will come up heads exactly eight times? Use the continuity correction.

18.21 A manufacturing process produces silicon chips. One fifth of the chips produced are defective.

a) If you purchase ten chips, what is the chance that at least eight of them are not defective?

b) If you purchase 100 chips, what is the chance that at most 24 are defective?

18.22 A pedometer counts the number of steps you take by sensing the jolts when your foot hits the ground. Some joggers use pedometers to determine how far they have run. Suppose you measure your gait and determine that your step span is, on the average, three feet with an SD of half a foot. You jog until the pedometer tells you that you have gone 10,000 steps. Since your average step is three feet, you conclude that your run has taken you around 30,000 feet, or 5.68 miles (there are 5280 feet in a mile).

a) What is the chance that you have in fact gone at least 5.5 miles (29,040 feet)?

b) What is the chance that you have run between 29,900 feet (5.66 miles) and 30,100 feet (5.70 miles)?

18.23 In a city, 20% of the workers have incomes over $40,000 per year. If 1600 workers are chosen at random with replacement, what is the chance that between 180 and 196 of those chosen have incomes over $40,000 per year?

V. Solutions

18.1 a) from 25.5 up; b) from 10.5 up; c) from 100.5 up; d) from 24.5 up; e) from 9.5 up; f) from 99.5 up

18.2 a) from 24.5 down; b) from 9.5 down; c) from 99.5 down; d) from 25.5 down; e) from 10.5 down; f) from 100.5 down

18.3 a) 39.5 to 60.5; b) 40.5 to 59.5; c) 100.5 to 199.5; d) 99.5 to 200.5; e) 99.5 to 199.5; f) 100.5 to 200.5

18.4 a) 0; b) 10; c) 38% [42% with continuity correction, including endpoints]

Solutions

18.5 $EV = 5000$, $SE = 50$; a) 5050 or more $= 16\%$; b) 5500 or more $\ll 1\%$

18.6 50%

18.7 $A(0.125) = 10\%$.

18.8 c

18.9 a) $[400!/(44!356!)](1/10)^{44}(9/10)^{356} = 0.051$; b) about 5% to 6%

18.10 a) $A(3.6) = 100\%$; b) $1 - 2 \times (1/4)^3 = 97\%$

18.11 $[100!/(50!50!)] \times (1/2)^{100} = 0.01$

18.13 a) average of box $= 2$; SD of box $= 1$; $EV = 2 \times 100 = 200$; $SE = 1 \times \sqrt{100} = 10$; $(210-200)/10 = 1$ SD; $A(1) + (1/2)(100\% - A(1)) = 68\% + 16\% = 84\%$; b) $(1/2)(100\% - A(1.125)) = 13\%$; c) $(1/2)(A(1.5) + A(1.0)) = 77.5\%$

18.15 a) $[9!/(7!2!)] \times (4/5)^7(1/5)^2 + (9!/8!)(4/5)^8(1/5)+(4/5)^9 = 74\%$; b) $EV = 720$, $SD = 0.4$, $SE = 12$; answer $= 50\% + (1/2)A(1.67) = 95\%$

18.17 $EV = -\$21.05$, $SE = 20$; chance $= (1/2)(100\% - A(1.05)) = 14.5\%$

18.19 a) 84%; b) $(1/2)A(1.5) = 43.5\%$

18.21 a) 68%; b) $50\% + (1/2)A(1.125) = 87\%$

18.23 $(1/2)A(1.0) = 34\%$

19

Sample Surveys

Example A: A survey is carried out by the planning department to determine the distribution of household size in a certain city. They draw a simple random sample of 1000 households. But after several visits, the interviewers find people at home in only 853 of the sample households. Rather than face such a high non-response rate, the planners draw a second batch of households, and use the first 147 completed interviews in the second batch to bring the sample up to its planned size of 1000 households. They count 3087 people in these 1000 households, and estimate the average household size in the city to be about 3.1 persons. Is this estimate likely to be too low, too high, or just about right? Explain.

Solution: This estimate is likely to be too high. With smaller households, the interviewer is less likely to find someone at home. So on the average, the survey procedure amounts to replacing smaller households with larger ones.

Example B: Ten percent of the families in a district have incomes over $25,000. A simple random sample of 900 families is taken from the district. The chance that there are exactly 90 families in the sample with incomes over $25,000 is closest to: a) 1%; b) 5%; c) 10%; d) 50%; e) 95%; f) 99%.

Solution: A sample size of 900 is enough for the normal histogram to approximate the histogram of the sum of the draws (see chapter 5). The box contains 10% ones (income over $25,000). The expected sum is 90 with an SE of 9. The chance of exactly 90 families in the sample having incomes over $25,000 is: $z = \dfrac{90.5 - 90}{9} = 0.05$; $A(0.05) = 4\%$.

So the answer is b) — closest to 5%.

Sample Surveys

Example C: A psychologist has a theory about what situations people find pleasant or unpleasant. To test it, he make up 25 items of the following kind:

Imagine that you like John, you like cats, and John likes cats. How pleasant is this situation?

Then in each item, the subject is asked whether he likes some person and some object, and if so, whether the person likes the object as well. This is called a situation of *cognitive consonance.* The psychologist now presents the 25 items to a volunteer subject, one item at a time. After each presentation, the subject rates the pleasantness of the situation on a scale of 1 to 5, with 1 standing for unpleasant, 3 for neutral, and 5 for pleasant. The average of the ratings is 4.2 with an SD of 0.9. The psychologist infers that with 95% confidence, this subject's long-run average response to this kind of item is in the interval from:

$$4.2 - 2.0 \times \frac{0.9}{\sqrt{25}} \text{ to } 4.2 + 2.0 \times \frac{0.9}{\sqrt{25}}$$

True, false, or silly? Explain.

Solution: Silly— the 25 items are not a random sample from all such items; the subject's response is heavily biased by many subjective attitudes and prejudices. It is difficult to see what box model is being used. Interview and response bias could both be large.

Example D: Why was a large-scale field trial needed to get convincing evidence of the effectiveness of the Salk Vaccine?

Solution: See section 1, Chapter 1.

179

20

Chance Errors in Sampling

I. Applying the Square Root Law to Percentages

The easiest way to answer a question about a percentage is to rephrase it as a question about the corresponding sum (or event), and use the EV and SE calculations described in Chapter 17.

Example A: A coin is tossed 100 times. What is the chance of getting between 45% and 55% heads?

Solution: "Between 45% and 55% heads" means "between 45 and 55 heads", The model is 100 draws from the box: $\boxed{0 \ \ 1}$; average = 0.5; SD = 0.5; EV = 50; SE = 5. So the chance is 68%.

Example B: A coin is tossed 400 times. What is the chance of getting between 45% and 55% heads?

Solution: "45% heads out of 400 times" means "180 heads", since 45% of 400 = 180. Likewise, if 55% of the 400 tossed are heads, this means you get 220 heads. So the question is, what is the chance of getting between 180 and 220 heads? The box model is 400 draws from: $\boxed{0 \ \ 1}$; average = 0.5; SD = 0.5; EV = 0.5 × 400 = 200; SE = $0.5 \times \sqrt{400}$ = 10. So the chance is A(z) = 95%.

Example C: What is the chance of getting less than 49% heads in 2500 tosses of a coin?

Solution: "Less than 49% heads" means "less than 1225 heads". EV = 1250 and SE = 25. So the chance is the area to the left of −1 under the normal, which is 16%.

Example D: A box contains 80% red tickets and 20% blue tickets. 400 tickets are drawn at random with replacement. What is the chance of getting more than 82% reds among the 400 drawn?

Solution: "More than 82%" means "more than 328", since 82% of 400 = 328. The box model is 400 draws from: $\boxed{1 \ 1 \ 1 \ 1 \ 0}$; average = 0.8; SD = 0.4; EV = 0.8 × 400 = 320; SE = $0.4 \times \sqrt{400}$ = 8. So the chance is the area to the right of +1 under the normal, which is 16%.

Computing the SE of a Percentage

Problems

20.1 The percentage of households owning more than one television in a certain city is 80%. 400 households are chosen at random with replacement. What is the chance that more than 82% of the households own more than one television?

20.2 400 draws are made from a box containing 10% red tickets and 90% black tickets. What is the chance of getting between 10% and 11.5% red tickets in the 400 drawn?

II. Computing the SE of a Percentage

The method of translating a question about a percentage into a question about a count, as presented in the last section, is good for working problems, since you do not have to learn any new procedures—you merely use what you learned in Chapter 17. However, if you want to get a better intuitive grip on the variability of percentages, you should learn to calculate the SE in percentage terms, not just in terms of the count. There are two ways to do this. One is:

1) Compute the SE of the count.
2) Convert this SE to a percentage by dividing by the number of draws.

Example A: What is the SE of the percentage of heads in 100 tosses of a coin?

Solution: From Example A in the last section, the SE of the *number* of heads is 5. So the SE of the *percentage* of heads is $5/100 = 5\%$. Thus you expect $50\% \pm 5\%$ heads in 100 tosses.

Example B: What is the SE of the percentage of heads in 400 tosses of a coin?

Solution: From Example B in the last section, the SE of the *number* of heads is 10. So the SE of the *percentage* of heads is $10/400 = 2.5\%$. Thus you can expect about $50\% \pm 2.5\%$ heads in 400 tosses.

Example C: What is the SE of the percentage of heads in 2500 tosses?

Solution: From Example C of the last section, the SE of the *number* of heads is 25. So the SE of the *percentage* of heads is $25/500 = 1\%$. Thus, in 2500 tosses you can expect to see about $50\% \pm 1\%$ heads.

Example D: A box contains 80% red tickets and 20% blue tickets. What is the SE of the percentage of red tickets in 400 random draws with replacement?

Solution: From Example D of the last section, the SE of the *number* of red tickets in 400 draws is 8. So the SE of the *percentage* of red tickets in the 400 draws is 8/400 = 2%. This tells you that you will get around 80% ± 2% red tickets in the sample.

Problems

20.3 In a city, 80% of the households own more than one television. A sample of size 400 is drawn at random with replacement from these households. What is the SE of the percentage of households in the sample with more than one television?

20.4 We make 400 random draws from a box containing 10% red tickets and 90% black tickets. What is the SE of the percentage of red tickets in the sample?

The second way to get the SE of a percentage is to use the formula. This is by no means necessary—it is merely convenient. It gives the same answer as the first method, just a little more directly. The formula for the SE of a percentage is:

$$SE = \frac{SD \text{ of box}}{\sqrt{\text{number of draws}}}$$

Notice that for this SE you *divide* by the square root of the number of draws. This is in contrast to the formula for the SE of a sum, where you *multiply* the SD of the box by the square root of the number of draws. If you have a hard time deciding which version to use, simply ignore the formula for percentages and always use the first method above—it will give you the same answer every time, with less confusion.

Example E: Use the formula to calculate the SE of the percentage of heads in: a) 100 tosses; b) 400 tosses; c) 2500 tosses; d) 10,000 tosses.

Solution: The box in each case is: $\boxed{0 \ 1}$; average = 0.5 and SD = 0.5. a) SE = $0.5/\sqrt{100}$ = 50%/10 = 5%; b) SE = $0.5/\sqrt{400}$ = 50%/20 = 2.5%; c) SE = $0.5/\sqrt{2500}$ = 50%/50 = 1%; d) SE = $0.5/\sqrt{10,000}$ = 50%/100 = 0.5

Notice that the answers to a), b) and c) are the same as those obtained in the corresponding example above. Notice also that the SE of the percentage goes down as the number of tosses goes up— this confirms the discussion about the variability of coins in Chapter 16.

The Correction Factor

III. The Correction Factor When Drawing Without Replacement

The square root formula for the SE is valid only for drawing *with* replacement (the EV formula is correct for drawing either with or without replacement). For drawing without replacement, as in most real world samples, you must multiply the SE by the "correction factor". The formula for the correction factor is:

$$\sqrt{\frac{\text{number of tickets in box} - \text{number of draws}}{\text{number of tickets in box} - 1}}$$

In terms of sampling from a population, the formula is written:

$$\sqrt{(\text{population size} - \text{sample size})/(\text{population size} - 1)}$$

Example A: Suppose a box has 1000 red and black tickets and 100 are drawn without replacement. What is the correction factor for the SE of the number of red tickets among the 100?

Solution: The proportion of red tickets in the box plays a role in computing the SE in the sample of 100, but it plays no role in computing the correction factor. The correction factor is:

$$\sqrt{(1000 - 100)/999} = \sqrt{900/999} = 0.95.$$

This tells you that the actual SE of the number of reds is 0.95 times the SE you get if you draw *with* replacement—it is smaller. So to compute the SE properly, you compute the SE for drawing with replacement using the ordinary SE formula, and then multiply that value by 0.95.

Problems

20.5 Compute the correction factor for each of the following samples.

	population size	sample size
a)	10,000	100
b)	10,000	400
c)	10,000	2,500
d)	10,000	8,100
e)	10,000	10,000
f)	100,000	100
g)	100,000	400
h)	100,000	2,500
i)	100,000	8,100
j)	100,000	10,000
k)	1,000,000	100
l)	1,000,000	400
m)	1,000,000	2,500
n)	1,000,000	8,100
o)	1,000,000	10,000

20.6 One of the correction factors in the last problem was zero. The correction factor times the SE with replacement equals the SE without replacement; so if the correction factor is zero the SE without replacement must be 0. Explain how this can be.

20.7 In which case does the correction factor make a larger correction to the SE?

i)   A: population size = 1,000,000; sample size = 400
      B: population size = 10,000; sample size = 400

ii)   A: population size = 10,000; sample size = 2500
      B: population size = 10,000; sample size = 100

iii) A: population size = 1,000,000; sample size = 10,000
      B: population size = 10,000; sample size = 100

20.8 In which case (in the problem above) does the sample size seem to have a greater effect on the correction factor (and thus the SE): the case where the population size is 10,000 or the case where it is 1,000,000?

IV. To Compute the SE When Drawing Without Replacement

1) Compute the SE as if you were drawing with replacement.

2) Compute the correction factor.

3) Multiply the "with-replacement SE" by the correction factor.

Example A: What is the SE of the number of reds in 100 draws without replacement from a box with 800 reds and 200 blacks?
Solution: The box is: $\lfloor$ 800 (1) 200 (0) $\rfloor$; average = 0.8; SD = 0.4.

1) The SE for the number of reds if you were drawing with replacement would be $0.4 \times \sqrt{100} = 4$.

2) The correction factor is equal to 0.94.

3) So the SE of the number of reds in this sample of 100 is slightly less than 4, namely $0.94 \times 4 = 3.7$.

Example B: A simple random sample of size 2500 is drawn from a population of 100,000 men and women. Exactly 50% of the population are men. What is the standard error of the number of men in the sample?
Solution: The box is: $\lfloor 0 \ 1 \rfloor$; average = 0.5; SD = 0.5.

1) Drawing with replacement, the SE of the number of men would be: $0.5 \times \sqrt{2500} = 25$

2) The correction factor is: $\sqrt{(100,000 - 2,500)/99,999} = 0.987$

3) So the SE drawing without replacement is $25 \times 0.987 = 24.686$.

184

Exercises

Example C: In Example B, what is the SE of the percentage of men in the sample?

Solution: 1) Drawing with replacement, the SE of the percentage of men would be $25/2500 = 1\%$. 2) The correction factor is still 0.987. 3) The SE of percentage of men $= 0.987 \times 1\% = 0.987\%$.

Problems

20.9 A box contains 10,000 tickets: 4000 reds and 6000 blacks. A sample of size 100 is drawn without replacement. What is the SE of the percentage of reds in the sample?

20.10 Looking at the table of correction factors in the solution to the first problem in the above section (20.5), in which situation does the sample size have a greater effect on the SE, the one with population size 10,000 or the one with population size 1,000,000?

20.11 A sample of size 2500 (without replacement) is taken in two separate situations. In each case, half the population are men and half are women. In the first case, the population size is 1,000,000 and in the second case it is 10,000,000. In which case is the SE of the percentage of men in the sample smaller?

V. Exercises

20.12 One hospital has 112 live births during the month of June, another has 432. Which is likelier to have 55% or more male births? Or are the chances equal? Explain (Note: there is about a 52% chance for a liveborn infant to be male).

20.13 A simple random sample of 15 classes is taken at a large university. All the students in these classes are interviewed. There are about 400 of them, and out of these 112, or 28%, are seniors. True/False: an estimate of the standard error for this percentage is 2.25%.

20.14 In 1971, women received 90% of the degrees in nursing, but only 9% of the MD's. These figures are for the entire US. Suppose that a certain medical society had 10,000 members. It now wants to choose a sample of 900 of them at random for a study. From its records, it is known that 10% of the members are women. What is the chance that between 7% and 13% of the subjects in the sample will be women?

20.15 A certain town is known to have 100,000 households with an average of 1.8 cars for each household. The SD is 0.6, and 80% of the households have at least one car. As part of a traffic survey, investigators

185

plan to take a sample of 1600 households. Find the chance that in their sample, over 81% of the households will have at least one car.

20.16 A poll is taken in a city of population 200,000. A simple random sample of size 2000 is chosen and polled. Another poll is to be taken in the same way in a second city of population 400,000. In order to obtain the same accuracy as in the first city, the sample size in the second city should be: a) 1000; b) 2000; c) 4000; d) 8000.

20.17 In a certain precinct, 80% of the voters are Republican. A simple random sample of size 400 is drawn (with replacement). Each person in the sample is polled and the percentage of Republicans in the sample is calculated. What is the chance that this percentage is between 78% and 88%?

20.18 As part of a new testing program, a state school system takes a simple random sample of 100 third grade classes, and gives a reading test to all the children in the sampled classes. There are 1400 third grade classes in the system and the average class size is 25.6 pupils with an SD of 3.0. How likely is it that at least 2500 children are tested in this part of the program?

20.19 The population of San Francisco is about 800,000. The population of Berkeley is about 100,000. In order to estimate the proportion of Democrats within each city, a random sample of 100 voters is taken in each. Each sample is found to contain the same number of Democrats. The standard error of the San Francisco estimate will be about: a) the same as the SE of the Berkeley estimate; b) 1/10 that of the SE of the Berkeley estimate; c) $1/\sqrt{8}$ that of the SE of the Berkeley estimate; e) cannot compare the SE's of the two estimates unless we are given the (common) number of Democrats in each sample.

20.20 A sample survey company advertises that they can determine the percentage of people in any category you want, within an SE of 1%, using a sample of only 2500 randomly chosen people.

a) Since the SE depends on the true proportion, which is unknown beforehand, how can they be certain 2500 is enough, particularly if the population in question includes several hundred million people?

b) If the town you wanted to sample had a population of 12,500, what kind of accuracy would the survey company get with a sample of 2500?

Solutions

20.21 The Census Bureau is planning to take a sample amounting to 1/2 of 1% of the population in each state in order to estimate the percentage of the population in that state with more than 12 years of education. Other things being equal, which of the following are correct?

a) The accuracy to be expected in New York (population 20 million) is about the same as the accuracy in Montana (population 1/2 million).

b) The accuracy to be expected in New York is quite a bit higher than in Montana.

c) The accuracy to be expected in New York is quite a bit lower than in Montana.

20.22 In 1972, the freshman class at UCB numbered about 3,000. About 20% of these students had a GPA of 3.4 or better. As part of a study, a sample of 400 freshmen is selected at random from the 3,000. The chance that the percentage in the sample with a GPA of 3.4 or more will be within 4% of the corresponding percentage among the 3,000 is:
a) at most 25%; b) at least 75%; c) about 95%.

VI. Solutions

20.1 This is identical to the box model experiment described in Example D—the answer is 16%.

20.2 "Between 10% ad 11.5%" means "between 40 and 46," since 10% of 400 = 40 and 11.5% of 400 = 46. The box model is 400 draws from: $\boxed{1\ 0\ 0\ 0\ 0\ 0\ 0\ 0\ 0\ 0}$.; average = 0.1; SD = 0.3; EV = 40 and SE = .6. So the chance is the area between zero and 1.0 under the normal, which is 34%.

20.3 This is the same problem as Example D—the SE is 2%. So you expect to get within about 2% of 80% in the sample with more than one television.

20.4 From 20.2 in the previous section, the SE of the *number* of red tickets in the sample is 6. So the SE of the *percentage* of red tickets in the sample is $6/400 = 1.5\%$. The sample will have about $10\% \pm 1.5\%$ red tickets in it.

20.5

	population size	sample size	correction factor
a)	10,000	100	.99504
b)	10,000	400	.97984
c)	10,000	2,500	.86607
d)	10,000	8,100	.43591
e)	10,000	10,000	0
f)	100,000	100	.9995
g)	100,000	400	.998
h)	100,000	2,500	.98743
i)	100,000	8,100	.95865
j)	100,000	10,000	.94869
k)	1,000,000	100	.99995
l)	1,000,000	400	.9998
m)	1,000,000	2,500	.99875
n)	1,000,000	8,100	.99594
o)	1,000,000	10,000	.99499

20.6 If the population size is 10,000 and the sample size is also 10,000, and you are drawing without replacement, the sample will contain the whole population. So there will be no variability in the sample—it will be the same every time you perform the experiment—and therefore the same percentage of what you are counting (red tickets men, etc.) will always be exactly equal to the percentage in the population. So the SE will be zero, as multiplying by the correction factor tells you.

20.7 i) B—notice that in B the sample size is larger relative to the population size than in A. Varying the sample size does not seem to affect the correction factor much at all (at least in the range of sample sizes tabulated)—they are all over 0.99, which is almost 1.0; ii) A—notice again that in A the sample constitutes a greater proportion of the population; iii) A—the two correction factors are about the same—notice that 100 out of 10,000 is 1%, and so is 10,000 out of 1,000,000.

20.8 10,000. With population size = 1,000,000, all the correction factors are greater than 0.99, and they vary only slightly with sample size.

20.9 The model is: 100 draws from the box: $\lfloor$ 4000 (1) 6000 (0) $\rfloor$; average = 0.4; SD = 0.5.

1) With replacement, SE (of percent reds) = $0.5/\sqrt{100}$ = 5%

2) Correction factor = $\sqrt{9900/9999}$ = 0.99504

3) Without replacement, SE = $0.99504 \times 5\%$ = 4.98%

Solutions

20.10 In the cases where the population size is 10,000 the correction factors vary much more dramatically with changes in the sample size—so the "without-replacement SE" changes more with sample size if the population size is 10,000 than if the population size is 1,000,000.

20.11 In the first situation the box model is: 2500 draws from box = $\lfloor$ 500,000 (0) 500,000 (1) $\rfloor$; average = 0.5; SD = 0.5; EV = 1250; with-replacement SE = 25/2500 = 1%, without-replacement SE = 1% × 0.99875 = 0.99875%. In the second situation, the box model is: 2500 draws from box = $\lfloor$ 5,000,000 (0) 5,000,000 (1) $\rfloor$; average = 0.5; SD = 0.5; EV = 1250; with-replacement SE = 1%; and without-replacement SE = 1% × 0.999999 = 0.999999%. So, the SE in the first situation is slightly smaller, but not by much—both SE's are equal almost exactly to 1%, the same value you would get if you were drawing with replacement. This shows that when the population size is large relative to the sample size, drawing without replacement is almost like drawing with replacement, and the SE depends mainly on the sample size.

20.13 False—the sample is a form of cluster sample. The SE of 2.25% would apply only if the sample was a simple random sample of 400 from the underlying 0–1 box.

20.15 We use a 0–1 box. The number of sample families owning at least one car is like the sum of 1600 draws from the box: $\lfloor$ 1 1 1 1 0 $\rfloor$. The percentage of 1's among the draws will be 80% ± 1% or so, so the chance is 16%.

20.17 We make 400 random draws random with replacement form a 0–1 box containing 80% 1's. The expected percentage of 1's in the sample is 80%, with an SE of 2%. The chance of between 78% and 88% 1's in the sample is, therefore: $(1/2)[A(1) + A(4)] = (1/2)[68 + 100] = 84\%$.

20.19 a)—see section 3, Chapter 20.

20.21 b) is right—the size of the sample will be much larger in New York, which is what counts.

21

The Accuracy of Percentages

I. An Estimate of the SE When the Contents of the Box Are Unknown

In any true sampling situation, the sampler does not know the composition of the population, i.e., what's in the box. If the box is a 0–1 box (Yes-No, Republican-Democrat, etc.), the sampler really needs to know only one thing about the box in order to calculate the SE of the sample percentage: the SD. Strictly speaking, the SD is not knowable without knowing the population percentage the sampler wants to estimate. In this situation, the sampler uses the following technique:

1) Look at the sample and compute percentage of "1's" in the sample.

2) Compute the SD of a box which has this percentage of "1's" in it.

3) In the formula for the SE, use this SD as the SD of the population box. This results in an *estimated* SE.

Thus the sample uses the sample SD as an estimate of the population SD.

Example A: A poll samples 400 people at random from a large population and finds 120 college graduates in the sample. Estimate the SE of the sample percentage.

Solution: 1) 120/400 is 30%. Assume that the population consists of 30% college graduates and 70% others. (You should understand that the population may have a very different percentage of college graduates—this assumption is made for the purpose of computing the SE only, and after computing the SD we will drop the assumption.)

2) The SD of a box with 30% "1's" and 70% "0's" is 0.46.

3) In 400 draws without replacement from a box with 30% "1's" and 70% "0's", the SE of the percentage of "1's" in the sample is: $0.46 \times \sqrt{400}/400 = 9.2/400 = 2.3\%$ (since the population is large, assume the correction factor $= 1.0$). So the estimated SE is 2.3%.

To Construct an Approximate X-% Confidence Interval

Problems

21.1 A simple random sample of size 2500 is drawn from the population of all US voters to determine the percentage who favor unilateral disarmament. Five hundred from the sample say they favor it. Estimate the SE of the sample percentage of those favoring disarmament.

21.2 In Example A above, assume the population percentage was really 35%, not 30% as suggested by the sample. In that case, what would be the true SE of the sample percentage?

21.3 Suppose in Example A the population percentage was really 50%. Then what would be the true SE of the sample percentage?

II. To Construct an Approximate X-% Confidence Interval

1) Compute the sample percentages.

2) Compute the estimated SE of the sample percentage, as outlined in the last section.

3) Using this as if it were the actual SE, subtract one SE from the sample percentage to get a lower value and add one SE to the sample percentage to get an upper value—the interval from the lower value to the upper value is an approximate 68%-confidence interval for the sample percentage. To get an approximate 95%-confidence interval, subtract 2 SE's from, and add 2 SE's to the sample percentage. To get an approximate 99%-confidence interval, do the same with 3 SE's.

Example A: In the situation described in Example A in the previous section, find a: a) 68%-confidence interval; b) 95%-confidence interval; c) 99%-confidence interval; for the sample percentage.
Solution: 1) The sample percentage is $120/400 = 30\%$.

2) The estimated SE is 2.3%, as seen in the example.

3) a) an approximate 68%-confidence interval for the sample percentage is from $30\% - 2.3\% = 27.7\%$ to $30\% + 2.3\% = 32.3\%$.

b) an approximate 95%-confidence interval for the sample percentage is from $30\% - 2 \times 2.3\% = 25.4\%$, to $30\% + 2 \times 2.3\% = 34.6\%$.

c) an approximate 99%-confidence interval for the sample percentage is from $30\% - 3 \times 2.3\% = 23.1\%$, to $30\% + 3 \times 2.3\% = 36.9\%$.

Problems

21.4 The 1976 Gallup Poll Pre-election Survey estimated the percent voting Democratic as 49.5% based on a sample of 3439. A standard error is attached to this estimate by the following procedure. The SE for the number of Democrats in the sample is given by: $\sqrt{3439} \times \sqrt{0.495 \times 0.505}$ = 29.32. Therefore, the SE for the percent Democratic is given by: 92.32/3439, which is about 0.9. Comment.

21.5 Los Angeles has about four times as many registered voters as San Diego. A simple random sample of registered voters is taken by each city to estimate the percentage who will vote for school bonds. Other things being equal, a sample of 4,000 taken in Los Angeles will be about: a) four times as accurate; b) twice as accurate; c) as accurate; as a sample of 1,000 in San Diego. Why?

21.6 There are about 25,000 high schools in the United States. Each high school has a principal. As part of a national survey of education, a simple random sample of 225 high schools is chosen. In 202 of the sample high schools the principal has an advanced degree.

a) If possible, find an approximate 95% confidence interval for the percentage of all 25,000 high school principals who have advanced degrees. If this is impossible, explain why.

The 25,000 high schools in the United States employ a total of about one million teachers. As it turned out, the 225 sampled high schools employed a total of 10,000 teachers, of whom 5,010 had advanced degrees.

b) If possible, find an approximate 95% confidence interval for the percentage of all one million teachers with advanced degrees. If this is impossible, explain why.

21.7 In a simple random sample of 900 students at a major state university, only 180 favor a return to the semester system. Find a 95% confidence interval for the percentage of students in the university favoring a return to the semester system.

III. Exercises

21.8 In a certain large university there happen to be exactly the same number of men as women. If a person were to take a simple random sample of the students at this university, how large a sample size would one have to take to be 68% confident that the proportion of women observed in the sample would be between 48% and 52%?

Solutions

21.9 We are interested in the proportion of the U.S. population which favors the death penalty. We are going to take a random sample of the population. Assume that the appropriate function of our sample observations has approximately a normal distribution. Approximately how large a sample should we take to be within 0.04 of the correct value of the proportion with probability 0.95, if that proportion is about 0.50?

21.10 In a simple random sample of 400 Berkeley residents taken by a polling organization, only 30% expressed support for the mayor. Find a 95% confidence level for the corresponding percentage in the whole population of Berkeley.

21.11 In a random sample of 1,000 voters, 530 favor Proposition 13. Find a 95%-confidence interval for the population percentage.

IV. Solutions

21.1 1) $500/2500 = 20\%$; 2) A box with 20% 1's and 80% 0's has SD $= 0.4$; 3) Estimated SE $= 0.4 \times \sqrt{2500}/2500 = 0.8\%$.

21.2 Box SD $= \sqrt{(.35)(.65)} = 0.48$. The SE of the sample percentage would be $0.48 \times \sqrt{400}/400 = 2.4\%$, not the 2.3% estimated by the sampler using the above technique. Note that the estimate of 2.3% is very close to the true value of 2.4%.

21.3 Box SD $= 0.5\%$. The SE of the sample percentage would be: $0.5 \times \sqrt{400}/400 = 2.5\%$. Note that even in this case, where the sample has given a very misleading idea about the population percentage, the estimate SE of 2.3% is still quite close to the true value of 2.5%.

21.4 This procedure does not apply to "multistage cluster samples" used by Gallup et al. The method works only for simple random samples from 0−1 boxes—see section 4 of Chapter 19 and sections 4 and 5 of Chapter 21 in FPP.

21.5 b)—each city corresponds to a 0−1 box, where 1 indicates the voter favors school bonds. By "other things being equal" we mean the percentage of 1's in the LA box equals the percentage of 1's in the San Diego box. So the SD's of the two boxes will be the same. The SE's of the two percentage estimates will be: SE of Los Angeles $= SD/\sqrt{4000}$; SE of San Diego $= SD/\sqrt{1000}$; so the SE of Los Angeles is half the SE of San Diego. In relative terms, the error gets smaller as the sample size increases.

193

21.6 a) This is possible, if we assume that the true percentage of high school principals who have an advanced degree is close to the observed percentage of 90% (so that the histogram for the sum of 225 draws follows a normal curve—see Chapter 18, section 5 of FPP for equivalent example with a biased coin).

The true percentage is unknown, but is estimated as 90%. The SD of the box is estimated as $\sqrt{0.90 \times 0.10} = 0.30$, the SE of the percentage estimate is therefore 2%. An approximate 95% confidence interval is 90% $\pm 4\%$.

b) This is not possible since the sample of 10,000 teachers is a cluster sample. We are unable to estimate an SE.

21.7 The true percentage of students favoring a return to the semester system is unknown but can be estimated as $(180/900) \times 100 = 20\%$. $\sqrt{0.20 \times 0.80} = 0.40$. The SE of the percentage is 1.3%. A 95% confidence interval is: 20% $\pm 2 \times 1.3\% = 17.4\%$ to 22.6%.

21.9 The distribution of sample percentages follows a normal distribution with an average of 50% (if the 0–1 box contains 50% 1's) and an SE given by: $SD \times 100/\sqrt{\text{number of draws}}$, where $SD = 0.50$. We require 50% $\pm 4\%$ to have a 95% chance of occurring. So 4% $= 2 \times SE$, so SE $= 2\%$, hence a sample of size 625.

21.11 The sample is a simple random sample from the whole voting population represented by the 0–1 box: $\boxed{\ ??\ \ (0)\quad ??\quad (1)\ }$ The percentage of 1's in the box is estimated as 53% with an SD of $\sqrt{0.53 \times 0.47} = 0.50$. The SE $= 1.6\%$. We estimate the percentage of 1's in the box as 53%, give or take 1.6%. A 95%-confidence interval is from 53% $\pm 2 \times 1.6\% = 49.8\%$ to 56.2%.

22

Measuring Employment and Unemployment

Example A: With the U.S. Census, why is it desirable in some instances to take a 5% sample as opposed to a complete census? What does the 5% sample lose?

Solution: Greater accuracy of survey data, because of improved quality control, over a complete census, in which mistakes are inevitable. See sections 6 and 7 of Chapter 22 in FPP.

23

The Accuracy of Averages

I. Computing the EV of the Average of Draws

If you draw a number of times from a box, with or without replacement, and compute the average of the numbers drawn, the answer will depend upon what is drawn—it is random. Its expected value, however, is just the average of the numbers in the box.

Example A: Thirty draws are made with replacement from the box: $\boxed{1\ 5\ 6}$. The average of these 30 numbers is computed. What is the EV?

Solution: The average of these 30 random draws is the average of the box, which is 4. You can understand this intuitively in several ways. For one, the 30 draws should fairly well have the same composition of values as you find in the box—in 30 draws you should get about ten 1's, ten 5's and ten 6's. So the sum should be about $(10 \times 1) + (10 \times 5) + (10 \times 6)$, and the average should be about:

$$\frac{(10 \times 1) + (10 \times 5) + (10 \times 6)}{30} = \frac{1 + 5 + 6}{3}$$

which is the average of the box. Equivalently, the sum of 30 draws should be about the EV of the sum, which is $30 \times$ (average of box), and so the EV of the *average* of these 30 draws should be about the EV of the sum divided by 30 which is: $30 \times$ (average of box)$/30 =$ average of box.

Problem

23.1 A large group of students has average GPA = 3.5. A sample of size 20 is drawn at random from this group and the average GPA in the sample is computed. What is the EV of the GPA in the sample?

Computing the SE of the Average of Draws

II. Computing the SE of the Average of Draws

The SE of the average tells you how variable the average will be. The easy way to compute the SE of the average of draws is:

1) Compute the SE of the *sum* (using the correction factor if the draws are without replacement).

2) Divide this by the number of draws.

Example A: One hundred draws are made at random with replacement from the box: ⌊1 6⌋ and the average of the 100 draws is computed. What is the SE?

Solution: The box has average = 3.5 and SD = 2.5.

1) The SE of the sum of the draws is $2.5 \times \sqrt{100} = 25$.

2) The SE of the average of the draws is $25/100 = 0.25$.

This tells you that the average of the draws will be the EV, which is 3.5 plus or minus 0.25. So you can expect to see sample averages like 3.25, 2.75 3.00, etc. But an average as high as 5.0 would be very unlikely.

Problem

23.2 We make 1600 draws at random with replacement from the box: ⌊2 2 2 2 4 4 4 6 6 8⌋. What is the SE of the average of these 1600 draws?

Note that the average of these draws *could* be as large as 8 or as small as 2. What the SE is telling you is that it will probably not be anywhere near these extreme values—it will probably be very close to 4. The small SE will not allow the sample average to stray very far from what it is "supposed" to be. This is an important characteristic of the sample average. If the sample size is reasonably large (1600 in this instance), the sample average and the box average are not very far from one another. This means that, by looking only at the average of the 1600 draws, you can get a fairly good idea of the average of the box. In most problems you are told the contents of the box, but in the real world you often do not know its contents and you wish to discover them. The stability of the sample average makes it possible to discover, with a good deal of accuracy, one important aspect of these contents: the average value. Another equivalent way to compute the SE of the average of draws is to use the formula:

SE of average $= (\text{SD of box})/\sqrt{\text{number of draws}}$.

Again, it is not necessary to use this formula. If you feel more comfortable with the first method, use it. The advantage of understanding the formula is that you can see clearly how the variability of the average declines as the number of draws goes up. If you increase the number of draws, the SD of the box is divided by a greater number and the SE of the average gets smaller. And doubling the number of draws does not cut the SE in half—there is a square root in the denominator. If you want to halve the SE, you have to increase the number of draws by four.

Example B: From the box in **23.2** above, a number of draws are made. What is the SE of the average of these draws if the number of draws is: a) 100; b) 400; c) 900; d) 1600; e) 2500.

Solution: a) $2/\sqrt{100} = 0.2$; b) $2/\sqrt{400} = 0.1$; c) $2/\sqrt{900} = 0.067$; d) $2/\sqrt{1600} = 0.05$; e) $2/\sqrt{2500} = 0.04$. Notice that you can decrease the SE from 0.2 (at 100 draws) to 0.1 by making 400 draws—300 more draws buy you twice as much accuracy (or half the SE). But to bring the SE down from 0.1 to half of that value, 0.05, you must make 1200 more draws; 1600 draws are needed. To bring the SE down further to half of 0.05, or 0.025, you must draw 6400 times. This is another feature of the sample average—you get a lot of accuracy right away, but subsequent improvements are progressively more costly in terms of the number of draws.

III. Confidence Intervals for the Average of the Box

This technique is the same one used in Chapter 21 for percentages. To compute an x-% confidence interval for the sample average:

1) Compute the SD of the numbers in the sample.

2) Using this SD as if it were the SD of the box, compute the estimated SE of the sample average.

3) A 68%-confidence interval for the box average is the interval starting at the sample average minus one SE and running up to the sample average plus one SE (sample average − SE, sample average + SE).

Example A: We make 1600 random draws from a box. The average of the 1600 draws is 5.3 and the SD is 0.2. What is a 68%-confidence interval for the average of the box?

Solution:

1) The SD = 0.2.

2) The estimated SE is: $0.2/\sqrt{1600} = 0.005$.

3) A 68%-confidence interval for the box average is the interval from 5.295 (= 5.3 − 0.005) to 5.305 (= 5.3 + 0.005).

Exercises

This is a very tight interval—due to the small sample SD and the large number of draws. In a loose sense, you can be "68% confident" that this interval contains the average of the box. More precisely, if you adopt the policy of constructing confidence intervals in this way, your intervals (which of course are random and depend upon the sample drawn from the box) will contain the average of the box about 68% of the time.

IV. Exercises

23.3 True/False: The chance that the box average in Example A is in the interval from 5.295 to 5.305 is equal to 68%.

23.4 In the situation in Example A, what is a 95%-confidence interval for the box average? A 99%-confidence interval?

23.5 For a batch of 100,000 tax forms, the reported gross income averages out to $12,000 with an SD of $6,000. Of these forms, 900 will be chosen at random for a detailed audit. What is the chance that the gross income reported on the forms chosen for audit will average over $11,800?

a) Over 90%; b) 80-90%; c) 20-80%; d) 10-20%; e) Under 10%

23.6 A sample of 2500 annual incomes is taken from a large population where the average annual income is X and the SD is $3,000 (from past data). The sample average is found to be equal to $10,000. Which of the following is true?

a) 95% of the incomes are between ($10,000 − $6,000) and ($10,000 + $6,000); b) 95% of the incomes are between X − $6,000 and X + $6,000; c) 95% of the incomes are between $10,000 − ($6,000/$\sqrt{2500}$) and $10,000 + ($6,000/$\sqrt{2500}$); d) in 95% of all possible samples the average X is between $10,000 − ($6,000/$\sqrt{2500}$) and $10,000 + ($6,000/$\sqrt{2500}$); e) in 95% of all possible samples the sample average will be between X − ($6,000/$\sqrt{2500}$) and X + ($6,000/$\sqrt{2500}$).

23.7 A population of 10,000 has an average of 203 with a SD of 30.0. A simple random sample of size 100 is taken. What is the probability that the average of the sample is less than 200 if:

a) no correction for finite population is made?

b) a correction for finite population is made?

V. Solutions

23.1 The box from which you are drawing has average = 3.5. So the EV of the average of the draws is 3.5.

23.2 The box average = 4 and SD = 2. 1) The SE of the sum of the 1600 draws is $2 \times \sqrt{1600} = 80$.

2) The SE of the average of the 1600 draws is $80/1600 = 0.05$.

The EV of the average of the 1600 draws is the average of the box, which is 4, so you expect the sample average to be about 4 ± 0.05.

23.4 False—the box average is either in that interval or it isn't. The 68% can be interpreted as a chance only if you imagine constructing many 68%-confidence intervals—in this use they will cover the box average about 68% of the time.

23.5 b

23.7 a) The SE of sample average is 3.0. The EV is 203 and sample average follows a normal distribution. Answer is 16%.

b) The SE of sample average is $\sqrt{(10,000 - 100)/9,999} \times 3.0 = 2.99$; the answer is 16%.

24

A Model for Measurement Error

I. Exercises

24.1 Six measurements of a quantity yielded the following observations: 1, 3, 4, 4, 5, 7 (unrealistic but picked in order to simplify calculation). Assuming the Gauss model and assuming that 2.57 is the 95 percent point from a t-curve with 5 degrees of freedom, a 95 percent confidence interval for the long run average would be:

a) $4 \pm 2\sqrt{10}/3$; b) $4 \pm 2\sqrt{10}/18$; c) $4 \pm 2.57 \times 2/\sqrt{6}$;

d) $4 \pm 2.57 \times \sqrt{10}/3$; e) can't compute because no population SD given.

24.2 True/False: Provided that an estimator for some parameter follows a normal curve, the 95 percent confidence interval for the parameter is: estimator $\pm 2 \times$ [SE (estimator)]

24.3 To produce a 95 percent confidence interval of the form: sample average ± 0.2 for the long-run average (population mean) of measurements which follow a Gauss model where the SD of the error box is 2, the sample size must be about:

a) 400; b) 100; c) 25; d) 1; e) can't tell without knowing sample average

24.4 True/False: Of 1,000 95% confidence intervals for the average, you would expect that about 950 would actually cover the average.

24.5 An investigator looks up the rainfall in a certain city on January 15 for the past 70 years. She finds the average rainfall on that day to be 0.30 inches and the SD to be about 0.14 inches. Then she calculates:
$\sqrt{70} \times 0.14$ inches = 1.17 inches; (1.17 inches)/70 = 0.0167 inches;
3×0.0167 inches = 0.05 inches, and concludes that the interval from 0.25 inches to 0.35 inches is a 99.7 percent confidence interval for the average rainfall on January 15 in the city. Is this conclusion justified? Why or why not?

24.6 Which of the following is not needed to compute a confidence interval for the population average?

a) Sample size; b) Confidence level; c) Population size; d) Sample average; e) Sample SD

24.7 True/False: If you have worked out a 95 percent confidence interval for the population average using the average of a sample of size 1600 and you would like to have a new interval with half the width but still with 95 percent confidence, you should increase the sample size to 6,400.

24.8 Suppose from the population of 18-year-old males on the Berkeley campus a random sample of size 49 is picked, and each is weighed. Suppose also that for this sample, the average weight is x = 160 pounds, with an SD of s = 4 pounds. Find a 98% confidence interval for the average weight of the population.

24.9 True/False: If a 95 percent confidence interval for the average strength of a certain type of wood is 12-15 psi, then one can conclude that in a sample of 100 wood beams approximately 95 would have strength between 12 and 15 psi. (psi = pounds per square inch.)

24.10 In 1976, the average of the daily maximum temperature at San Francisco Airport was 64.2 degrees, with an SD of 7 degrees. Now, $\sqrt{365} \times (7 \text{ degrees}) \approx 134$ degrees; (134 degrees)/365 $\approx$ 0.37 degrees. True/False: a 95 percent confidence interval for the average daily maximum temperature at San Francisco Airport is 64.2 ± 0.74 degrees.

24.11 A new penny is weighed 25 times on a sensitive scale with digital readout. The 25 weights are recorded, and their average and SD calculated. These are 3.10 and 0.15 grams respectively. Find a 95 percent confidence interval for the weight of the penny. The coin is about to be weighed once more. With chance about 95 percent, what interval, in grams, will the readout show?

24.12 The speed of light is measured 2,500 times by a new process. The average of these 2,500 measurement is 299,774 kilometers per second, with an SD of 14 kilometers/sec. Assuming the Gauss model, a 99.7 percent confidence interval for the long-run average value of the speed of light by this measurement process is:

a) $299,774 \pm 3 \times 14$ kilometers/sec.; b) $299,774 \pm 3 \times 14 \times \sqrt{2500}$ kilometers/sec.; c) $299,774 \pm 3 \times 14/\sqrt{2500}$ kilometers/sec.;
d) $299,774 \pm 3 \times 14 \times \sqrt{2500}$ kilometers/sec.; e) $299,774 \pm 3 \times 14/\sqrt{2500}$ kilometers/sec.

Solutions

24.13 The speed of light is measured 2,500 times by a new process. The average of these 2,500 measurement is 299,774 kilometers per second, with an SD of 14 kilometers per second.

a) Find an approximate 95 percent confidence interval for the speed of light. (You may assume the Gauss model, with no bias. Show your work.)

b) Now the investigators determine the speed of light once more by the same procedure, and get 299,781 kilometers per second. It this a surprising result? Why or why not?

II. Solutions

24.1 c—see section 6, Chapter 24 of FPP for discussion of use of SD and student t-critical value.

24.3 a—SE of the average $= 2.0/\sqrt{400} = 0.1$, then use the sample average $\pm\, 2 \times$ (SE of average).

24.5 No—the sample is not a simple random sample of the rainfall on January 15.

24.7 True—assuming same estimate of SD of average.

24.9 False—the confidence interval is for the population average and says nothing about the distribution of values in the population.

24.11 Each measurement is equal to the exact weight plus a draw from the error box. Bias is assumed to be zero. The SD of the box is unknown, but is estimated as 0.15g. The distribution is about normal (t with 24 degrees of freedom is close to normal). The confidence interval is 3.10 ± 0.06g.

24.13 As in the previous problem, we assume the error box is normal. Otherwise a student's t-curve should be used. a) $299,774 \pm 4.80$ kilometers per second; b) No—the SD of the error box is estimated as 14 kilometers per second, so the observation looks to be within about one SD of the true average.

25

Chance Models in Genetics

I. Exercises

25.1 The seeds of Mendel's peas come in two colors: yellow and green. Seed color is controlled by one gene-pair, yellow being dominant and green recessive. Suppose that Mendel crossed a pure yellow plant with a pure green, getting a number of first-generation hybrid plants. Suppose he then crossed these first-generation hybrids with some pure green plants, obtaining 1,600 seeds. What is the chance that 850 or more of these seeds turn out to be yellow?

    a) about $850/1,600 = 53\%$; b) about $(1,600 - 850)/1,600 = 47\%$; c) about 5%; d) about 0.5%; e) none of these

25.2 Mendel crosses hybrid smooth peas with pure wrinkled peas, getting 2,500 progeny. From genetic theory we know that wrinkled is recessive and smooth dominant, so that he will get some wrinkled peas and some smooth peas among the 2,500. What is the expected number of wrinkled peas?

    a) 625; b) 1,000; c) 1,250; d) 1,500; e) 1,875; f) 2,000

25.3 In the preceding problem, what is the SE of the number of wrinkled peas among the 2,500?

    a) 25; b) 50; c) 75; d) 100; e) 125

25.4 In the preceding problems, what is the chance that Mendel will get exactly the expected number of wrinkled peas?

    a) 50%; b) 68%; c) 5%; d) Less than 1%

25.5 Mendel crosses hybrid smooth peas with hybrid smooth peas, getting 10,000 progeny. Given that wrinkled is recessive and smooth is dominant, what is the expected number of wrinkled peas among the 10,000 progeny?

    a) 1,000; b) 2,500; c) 5,000; d) 7,500; e) 9,000

Solutions

25.6 About what is the SE of number of wrinkled peas in the progeny?

a) 25; b) 45; c) 75; d) 100; e) 125

25.7 What is the chance that he will get exactly the expected number of wrinkled peas?

25.8 Eye color in humans is determined by one gene-pair, with brown dominant and blue recessive. In a certain family, the husband had a blue-eyed father; he himself has brown eyes. Likewise, the wife had a blue-eyed father, and she has brown eyes.

a) If they have two children, what is the chance both children will have brown eyes?

b) If they have four children, what is the chance that all four have brown eyes?

25.9 Yellow peas are recessive and green peas are dominant. In a series of experiments, pure green peas are crossed with first-generation hybrid peas to yield 1,600 peas in each experiment, about half of which are yellow and half of which are green. The number of green peas is counted in each experiment: 805 are obtained in the first, 792 in the second, 785 in the third, and so on. The list of these results is compiled and the SD of this list is computed. About how large should this SD be? Explain.

25.10 Snapdragons come in two basic colors, red and white. Each of a series of experiments crossing red with white yields 400 progeny. The number of red snapdragons yielded in each experiment is compiled into a list. The SD of this list is computed. About how large should the SD be? Explain.

II. Solutions

25.1 d—expect 800 yellows, SD of box = 0.5, so SE of sum = 20. 850 is 2.5 SE's above 800 and $A(2.5) = 99\%$. So chance of more than 850 yellows is $(1/2)(100 - 99) = 0.5\%$.

25.3 a—SD of box = $\sqrt{0.5 \times 0.5} = 0.5$ under Mendel's theory. The SE of sum = $\sqrt{2500} \times$ (SD of box) = 25.

25.5 b—25% chance of wrinkled peas

25.7 Less than 1%

25.9 About 20, the SE of the sum of 1600 draws from a 0–1 box containing 50% 1's.

205

26 and 27

Tests of Significance

I. Exercises

26.1 In the preceding five years, entering students at a certain university had an average SAT verbal score of 612 points. A simple random sample of 100 students is taken from this year's entering class. The average SAT verbal score for these students is 594 points, with an SD of 80 points. Does this show a decline in entering students' verbal abilities? Carry out the appropriate test of significance.

26.2 In an ESP study, a subject is asked to guess the outcome of many successive trials of a random experiment. The test statistic is computed and the observed significance level is 0.04. Assuming there are no flaws in the experiment design, which of the following statements are correct?

a) the probability that the null hypothesis is correct is 0.04; b) about 4% of all researchers who conduct a replication of this experiment on the same subject would obtain results this extreme; c) the chance the subject has ESP is 4%; d) the chance the subject does not have ESP is 4%; e) assuming the subject has ESP, the chance he would produce results this extreme is 4%; f) assuming the subject did not have ESP, the chance he would produce results this extreme is 4%; g) the subject definitely has statistically significant ESP; h) the subject definitely has ESP, but it is not statistically significant.

26.3 A large lecture course has 1000 students. On a midterm exam, the class average was 50 with an SD of 10. It is suggested that the early morning section has more serious students and would therefore have higher scores. This section has 25 students; their average score is 53.

a) Formulate the null hypothesis.

b) Compute z and P.

c) Can this higher average be reasonably explained by chance variation?

26.4 A gambler claims that when a certain coin is tossed, tails comes up too often. To test this assertion, the coin is tossed 900 times. It comes up heads 435 times. What do you conclude? Make a test.

Exercises

26.5 To test a genetic model, an experimenter sets up a null hypothesis, collects data and computes the z-statistic. A z-statistic of zero would be expected according to the null hypothesis, and the experimenter obtains a z-statistic of 1.65. Is each of the following statements true or false?

a) the chance the model is correct is about 95%; b) the chance the model is correct is about 5%; c) the chance of getting results this extreme again is about 95%; d) the chance of getting results this extreme again is about 5%; e) although these results are extreme, they cannot unambiguously refute the model, as there is always a chance of getting results that deviate from those predicted; f) the observed significance level is about 5%; g) the observed significance level is about 95%; h) if the null hypothesis is correct, the chance of getting a z-statistic this large or larger is about 5%.

26.6 With a perfectly balanced roulette wheel, in the long run, red numbers should turn up 18 times in 36. To test its wheel, one casino records the results of 380 plays, finding 214 red numbers.

a) Formulate the null and alternative hypotheses as statements about a box model.

b) Compute z and P.

26.7 Six hundred students took an introductory course in biology. On the final exam, the average grade was 70 with an SD of 25. One TA's section of 25 students had an average score of only 65. In defending himself against a charge of poor teaching, which of the following arguments should the TA use? There may be more than one correct answer.

a) the TA's students' grades were low because of the regression effect, not because of his teaching; b) the difference between the average of his section and that of the whole class can easily be explained by chance; c) it is well known that performance of students and teaching abilities of TA's are not correlated, so you would expect his students to do a little below average; d) none of these arguments make sense.

26.8 A new type of tire is given an endurance test to see whether or not it can average 20,000 miles under the worst conditions. A sample of 100 sets is tested and lasts an average of 20,226 miles, with an SD of 863 miles. The company does not wish to market the tires unless they are shown to wear significantly more than 20,000 miles under these conditions. Will they be marketed? A significance level of 0.01 is used.

26.9 True/False:

a) The larger the sample size, the more faith you have in a P-value calculated using a z-test based on the sample average.

b) In hypotheses testing if the P-value is 1 percent, there is a 1 percent chance that the null hypothesis being tested is true.

c) It is a good idea to look at your data to help you decide which hypothesis to test with it.

d) The P-value does not depend on the level of significance.

e) As the P-value gets smaller, the evidence against the null hypothesis gets stronger.

f) The smaller the P-value, the less likely the null hypothesis is true.

26.10 You have a hypothesis that the percentage of college-educated people is 90% and we want to test this against the alternative hypothesis that the percentage is less than 90%. You take a sample of size 25×25 = 625 and find that the sample p = 0.872. Assuming that your significance level is .05, which of the following should you do?

a) reject the null hypothesis because the z-statistic is less than -1.645

b) accept the null hypothesis because the z-statistic is more than -1.645

c) reject the null hypothesis because the z-statistic is less than -1.965

d) none of the above

II. Solutions

26.1 The SD of the box estimated as 80, so SE of average = 8. z = (594 - 612)/8 = -2.25, so p = 1%. Looks like a decline.

26.3 a) This is 25 draws at random without replacement from a box of 1000 tickets. Each ticket shows a score. The average of the box is 50 with an SD of 10. Null hypothesis: sample average of 53 is due to chance variation; b) z = (53 - 50)/2, p = 7%; c) Yes—it would happen by chance about 7% of the time.

26.5 False: a, b, c, d, g, i; True: e, f, h

26.7 Use b; z = -1, p = 16%.

26.9 a) True; b) False—see section 3, Chapter 26; c) False—you should know your null and alternative hypotheses before you ever design your experiment; d) True—it depends on the null/alternative hypothesis and the data; e) True—see section 3, Chapter 26; f) False—the p-value is found assuming the null is true.

28

The Chi-Square Test

I. Exercises

28.1 A random sample of 10,000 voters is taken and classified according to income status and political party affiliation. The following table summarizes the observations:

	Annual Income Below $20,000	Annual Income Above $20,000
Republican	3,960	1,040
Democrat	4,040	960

a) Calculate a 95 percent confidence interval for the percentage of voters in the population who have income exceeding $20,000.

b) Test the hypothesis that the percentage of voters who are Republicans *and* who have annual incomes exceeding $20,000 is 10 percent, against the alternative, that the percentage is greater than 10 percent.

c) Is party affiliation independent of income status?

28.2 A genetic model predicts plants of four different types from a given crossing. The types are A, B, C, D, and the model predicts 10% A, 20% B, 30% C, and 40% D. To test the model we examine 100 progeny of the crossing, and find that there are 15 A's, 25 B's, 30 C's and 30 D's.

a) Formulate the null hypothesis.
b) Compute the test statistic.
c) What is the observed significance level?

28.3 In the above experiment, which of the following statements are correct?

a) the model is incorrect with probability more than 95%; b) the model is correct with probability more than 95%; c) the chance of obtaining results this extreme again is less than 5%; d) the chance of obtaining results this extreme again is more than 5%; e) these results, no matter how extreme, cannot unambiguously refute the model, as there is always a chance of getting results that deviate from those expected under the model.

28.4 Suppose a researcher crosses pink snapdragons with other pink snapdragons. Genetic theory shows that 1/4 of the offspring should be red, 1/2 should be pink, and 1/4 should be white. The research reports that, among 800 offspring of such a crossing, 170 were red, 400 were pink, and 230 were white. Chi-square is used to test the credibility of the results, taking as the null hypothesis the assertion that these results are due to genetic variation, and as the alternative hypothesis that some effect has perturbed the results away from the expected proportions.

a) The chi square statistic is: i) 0.4; ii) 0.7; iii) 4.0; iv) 4.7; v) 9.0.

b) The degrees of freedom is: i) 1; ii) 2; iii) 3; iv) 4; v) 5.

c) The observed significance level is closest to: i) 1%; ii) 5%; iii) 30%; iv) 70%; v) 95%; vi) 99%.

28.5 An extra credit problem was given to a statistics class. The students choosing to do the problem were to collect 100 numbers from books, newspapers, almanacs, etc., and record the first digit in each number, obtaining a list of 100 digits. Then they compiled from this list a table of frequencies of the various digits, such as:

Digit	1	2	3	4	5	6	7	8	9
Number of Occurrences	45	22	10	6	8	3	3	1	2
Expected Number of Occurrences	57	17	9	6	4	3	2	1	1

The theoretical distribution of these numbers is known, and the expected number of 1's, 2's, ... is the last row of the above table.

a) To check whether a student actually drew up the digits from the sources as instructed (as opposed to making up 100 numbers), what kind of test should be used?

b) What is the value of the test statistic for the data above?

c) What is the observed significance level?

d) Out of 100 students who turned in the assignment, the data of four students failed the above test at the 5% level. What can you conclude about how these students collected their data?

Exercises

28.6 A die is rolled 120 times. Here are the results:

number	1	2	3	4	5	6
occurrence	13	21	17	25	14	30

a) Make a chi-square test to see if the die is fair. What do you conclude?

b) Make a z-test to see if even numbers come up too often. What do you conclude?

28.7 Plants with purely white flowers and others of the same genus with purely red flowers are crossed. The first generation is pink.

a) What are the colors and their proportions in the second generation?

b) Consider the experiment where the second generation had 141 white, 292 pink, and 132 red ones. Would you accept the null hypothesis at the 1% level?

28.8 To find out whether students should be required to take a statistics course, a campus poll is taken; the results are tabulated below.

	Freshmen	Sophomores	Juniors	Seniors	Totals
Should	99	98	98	97	392
Should not	121	112	92	83	408
Totals	220	210	190	180	800

Using the data given, test the hypothesis that responses do not differ according to class; that is, that the proportion in each category will be the same, regardless of class, in the total population of students. Use a 0.05 level of significance.

28.9 In a random sample of 300 shoppers, 120 prefer brand A, 90 prefer brand B, and 90 prefer brand C. Calculate chi-square for the hypothesis that the three brands are equally preferred in the population. What is the p-value for this chi-square?

28.10 True/False: A chi square test is to be used to test the hypothesis that five types of hair color (blonde, brown, black, red, other) occur in specified proportions in a large population. A random sample of size 100 is taken and classified according to hair color. The degrees of freedom for this test is 4.

28.11 A researcher plans to test the fairness of a roulette wheel. The results of 3,800 spins are recorded: 1,750 came up red, 1,800 came up black, and 250 came up green. (If the wheel is fair, the chance of red is 18/38, the chance of black is 18/38, and the chance of green is 2/38.)

a) Formulate the null hypothesis.

b) Compute the test statistic. It should include information from all three observed values.

c) What is the observed significance level?

II. Solutions

28.1 a) The confidence interval is $20\% \pm 2 \times 0.4\% = 20 \pm 0.8\%$;
b) $z = (10.4 - 10)/0.30 = 1.33$, $p = 9\%$, so accept the null hypothesis;

c)

	Expected	
Republicans	4000	1000
Democrats	4000	1000

Chi-square $= 4.0$ with 3 degree of freedom, $p < 5\%$. This argues against the null hypothesis of independence.

28.3 e

28.5 a) Chi-square test; b) Chi-square $= (45-57)^2/57 + \cdots + (2-1)^2/1 = 9.61$; c) $p = 30\%$; d) This is OK: even when the null hypothesis is true, we expect to get $p < 5\%$ about 5% of the time. So 4 out of 100 is OK.

28.7 a)

Colors	red	pink	white
Proportions (expected)	25%	50%	25%

b) Chi-square $=$
$$\frac{(141-141.25)^2}{141.25} + \frac{(292-282.5)^2}{282.5} + \frac{(132-141.25)^2}{141.25} = 0.93,$$
$p = 60\%$. Yes, accept the null hypothesis.

28.9 Chi-square $= 6.0$, $p = 5\%$

28.11 a) The wheel is fair; the fractions of "tickets" labeled red, black and green in the box are respectively, 18/38, 18/38 and 2/38. The difference is due to chance variation.

b) Chi-square $= \dfrac{(1750-1800)^2}{1800} + \dfrac{(1800-1800)^2}{1800} + \dfrac{(250-200)^2}{200} = 13.9$,
degrees of freedom $= 2$, c) $p < 1\%$